Springer Series in
Surface Sciences

Editor: Robert Gomer

Springer Series in **Surface Sciences**

Editors: G. Ertl, R. Gomer and D. L. Mills   Managing Editor: H. K. V. Lotsch

H.-J. Güntherodt R. Wiesendanger (Eds.)

Scanning Tunneling Microscopy I

General Principles and Applications to Clean and Absorbate-Covered Surfaces

Second Edition

With Contributions by
D. Anselmetti   R.J. Behm
P.J.M. van Bentum   S. Chiang   H.-J. Güntherodt
R.J. Hamers   H.J. Hug   H. van Kempen   Y. Kuk
H. Rohrer   R. Wiesendanger   J. Wintterlin

With 172 Figures

Springer-Verlag

Berlin Heidelberg New York
London Paris Tokyo
Hong Kong Barcelona
Budapest

Professor Dr. Hans-Joachim Güntherodt

Department of Physics, University of Basel, Klingelbergstrasse 82,
CH-4056 Basel, Switzerland

Professor Dr. Roland Wiesendanger

Institute of Applied Physics, Universität Hamburg, Jungiusstrasse 11,
D-20355 Hamburg, Germany

Series Editors

Professor Dr. Gerhard Ertl

Fritz-Haber-Institut der Max-Planck-Gesellschaft, Faradayweg 4–6,
D-14195 Berlin, Germany

Professor Robert Gomer, Ph.D.

The James Franck Institute, The University of Chicago, 5640 Ellis Avenue,
Chicago, IL 60637, USA

Professor Douglas L. Mills, Ph.D.

Department of Physics, University of California,
Irvine, CA 92717, USA

Managing Editor: Dr. Helmut K. V. Lotsch

Springer-Verlag, Tiergartenstrasse 17,
D-69121 Heidelberg, Germany

ISBN 3-540-58415-3 2. Auflage Springer-Verlag Berlin Heidelberg New York
ISBN 0-387-58415-3 2nd edition Springer-Verlag New York Berlin Heidelberg

ISBN 3-540-54308-2 1. Auflage Springer-Verlag Berlin Heidelberg New York
ISBN 0-387-54308-2 1st edition Springer-Verlag New York Berlin Heidelberg

Cip data applied for.

Typesetting: Macmillan India Ltd., India

SPIN 10478881   54/3140-5 4 3 2 1 0 – Printed on acid-free paper

Preface to the Second Edition

Since the first edition of "Scanning Tunneling Microscopy I" has been published, considerable progress has been made in the application of STM to the various classes of materials treated in this volume, most notably in the field of adsorbates and molecular systems. An update of the most recent developments will be given in an additional Chapter 9.

The editors would like to thank all the contributors who have supplied updating material, and those who have provided us with suggestions for further improvements. We also thank Springer-Verlag for the decision to publish this second edition in paperback, thereby making this book affordable for an even wider circle of readers.

Hamburg, July 1994   *R. Wiesendanger*

Preface to the First Edition

Since its invention in 1981 by G. Binnig, H. Rohrer and coworkers at the IBM Zürich Research Laboratory, scanning tunneling microscopy (STM) has developed into an invaluable surface analytical technique allowing the investigation of real-space surface structures at the atomic level. The conceptual simplicity of the STM technique is startling: bringing a sharp needle to within a few Ångstroms of the surface of a conducting sample and using the tunneling current, which flows on application of a bias voltage, to sense the atomic and electronic surface structure with atomic resolution! Prior to 1981 considerable scepticism existed as to the practicability of this approach.

In the past ten years the field of STM has grown rapidly and is still attracting researchers from various other fields. This fact is reflected by the ever increasing number of participants at STM conferences. STM's success is based on it being a powerful local probe, capable of imaging, measuring and manipulating matter down to the atomic scale in almost any environment: in air, in inert gas atmospheres, in liquids, in ultrahigh vacuum, and from low temperatures up to several hundred degrees centigrade. Due to its wide applicability, STM has been adopted by many different scientific disciplines including solid-state physics, materials research, chemistry, biology, and metrology. The scope has broadened even further since the development of related scanning techniques which are referred to as "SXM" techniques where "X" stands for any kind of interaction between a sharp probe tip and a sample surface. The most developed SXM technique – apart from STM – is scanning force microscopy (SFM), invented in 1986 by G. Binnig, C. F. Quate and Ch. Gerber at Stanford University. SFM also allows the study of surfaces of bulk insulators down to the atomic level – a field where almost all conventional surface analytical techniques have failed. The whole class of SXM techniques facilitates the study of a wide range of nanometer-scale surface properties.

The present book is the first of two volumes devoted to STM and related techniques. After an introduction to the field of STM (Chap. 1), H. Rohrer gives his personal historical view of the birth of STM (Chap. 2). The applications of STM covered by the first volume include metals (Chap. 3 by Y. Kuk and Chap. 4 by J. Wintterlin and R. J. Behm), semiconductors (Chap. 5 by R. J. Hamers), layered materials (Chap. 6 by R. Wiesendanger and D. Anselmetti), molecular imaging (Chap. 7 by S. Chiang) and superconductors (Chap. 8 by P. J. M. van Bentum and H. van Kempen). The second volume will include further applications of STM in fields such as electrochemistry, biology, and nanometer-scale surface modifications. It will also include the description and applications of related techniques. Finally we plan to present the detailed theory of SXM techniques in a third volume.

Our first two volumes on STM and related techniques are intended for researchers and scientists in the various disciplines including physics, chemistry, biology and metrology, and also for graduate students. These two volumes should be helpful to those who are active in the field of STM as well as non-experts who have become interested in these novel techniques. Therefore, it has been the aim to include many representative results together with comprehensive lists of citations to help the reader to navigate through the rapidly growing number of publications in the field of STM. Even undergraduate students or readers with little knowledge of the natural sciences should find the two volumes on STM exciting, with their description of a fascinating technique that allows us to visualize the beauty of nature on an atomic scale and the richness of structures at the submicron level. This aspect has surely played more than a chance role in stimulating the rapid development of the STM field.

It is a pleasure for the editors to thank all the authors who have contributed to the first two volumes on STM and related techniques. We also acknowledge the pleasant collaboration with Springer-Verlag. Finally, we all thank the inventors of STM and its relatives and also the manufacturers of the instruments for providing the foundation for the enormous development that has taken place during the past ten years. It is the hope of the editors that these two volumes on STM and related techniques will further stimulate both basic and applied research in this exciting field.

Basel, October 1991   *H.-J. Güntherodt*
                                                            R. Wiesendanger

Contents

Contributors

D. Anselmetti

Institut für Physik, Universität Basel, Klingelbergstrasse 82,
CH-4056 Basel, Switzerland

R. J. Behm

Abteilung Oberflächenchemie und Katalyse, Universität Ulm,
D-89069 Ulm, Germany

P. J. M. van Bentum

Research Institute for Materials, University of Nijmegen, Toernooiveld,
NL-6525 ED Nijmegen, The Netherlands

S. Chiang

IBM Research Division, Almaden Research Center, 650 Harry Rd.,
San Jose, CA 95120-6099, USA

H.-J. Güntherodt

Institut für Physik, Universität Basel, Klingelbergstrasse 82,
CH-4056 Basel, Switzerland

R. J. Hamers

Dept. of Chemistry, University of Wisconsin,
Madison, WI 53706, USA

H. J. Hug

Institut für Physik der Universität Basel, Klingelbergstrasse 82,
CH-4056 Basel, Schweiz

H. van Kempen

Research Institute for Materials, University of Nijmegen, Toernooiveld,
NL-6525 ED Nijmegen, The Netherlands

Y. Kuk

Department of Physics, Seoul National University, Seoul 151-742, Korea

H. Rohrer

IBM Zürich Research Laboratory, Säumerstrasse 4,
CH-8803 Rüschlikon, Switzerland

R. Wiesendanger

Institut für angewandte Physik, Jungiusstrasse 11,
D-20355 Hamburg

J. Wintterlin

Fritz-Haber-Institut der Max-Planck-Gesellschaft, Faradayweg 4−6,
D-14195 Berlin, Germany

1. Introduction

R. Wiesendanger and *H.-J. Güntherodt*

With 5 Figures

Since the first successful experiments by G. Binnig, H. Rohrer and coworkers at the IBM Zürich Research Laboratory in March 1981, scanning tunneling microscopy (STM) has developed into an invaluable and powerful surface and interface analysis technique. Some visions from the early days of STM have now been realised, and future applications may exist that can hardly be imagined at present. Only five years after the first successful operation of a scanning tunneling microscope (also abbreviated to STM), Binnig and Rohrer received the Nobel Prize in physics for 1986 together with Ruska for his contributions to the development of electron microscopy. In the following, we will briefly review the historical background of electron tunneling experiments. For a more comprehensive historical review of pre-microscope tunneling experiments, the reader is referred to the article by *Walmsley* [1.1].

1.1 Historical Remarks on Electron Tunneling

The phenomenon of tunneling has been known for more than sixty years – ever since the formulation of quantum mechanics. As one of the main consequences of quantum mechanics, a particle such as an electron, which can be described by a wave function, has a finite probability of entering a classically forbidden region. Consequently, the particle may tunnel through a potential barrier which separates two classically allowed regions. The tunneling probability was found to be exponentially dependent on the potential barrier width. Therefore the experimental observation of tunneling events is measurable only for barriers that are small enough. The concept of electron tunneling was first applied theoretically to problems such as the ionization of hydrogen atoms in a constant electric field [1.2], the dissociation of molecules [1.3], the field emission from metals in intense electric fields [1.4] and the contact resistance between two conductors, separated by an insulating layer [1.5–8].   •

Electron tunneling was observed experimentally in *p-n* junctions by *Esaki* [1.9, 10] and in planar metal–oxide–metal junctions by *Giaever* [1.11–17]. Tunneling of Cooper pairs between two superconductors was predicted by *Josephson* [1.18–20]. For their contributions to the investigation of electron tunneling phenomena, these authors received the Nobel Prize in physics for 1973. In his Nobel Prize lecture, Giaever explained why they used planar metal–oxide–metal tunneling junctions instead of the better defined metal–vacuum–metal junctions at that time: "To be able to measure a tunneling current the two metals must be spaced no more than about 100 Å apart, and we decided early in the game not to attempt to use air or vacuum between the two metals because of problems with vibration".

The first observation of metal–vacuum–metal tunneling was reported in 1971 by *Young* et al. [1.21–24]. Additionally, these authors developed a novel instrument, which they called a topografiner, for the measurement of the topography of metal surfaces. A field emission tip was scanned over a sample surface by means of two orthogonally mounted piezoelectric drives, while a third orthogonal piezoelectric drive, on which the field emission tip was mounted, was part of a feedback loop system keeping the field emission tip at a constant distance of several hundred ångstroms above the sample surface. The voltage which had to be applied to this third piezoelectric drive to ensure a constant distance between tip and sample surface, was used to measure the vertical position of the tip and therefore the topography of the sample surface. Due to the relatively large distance between tip and sample surface of several hundred ångstroms in the field emission regime, the topografiner achieved a

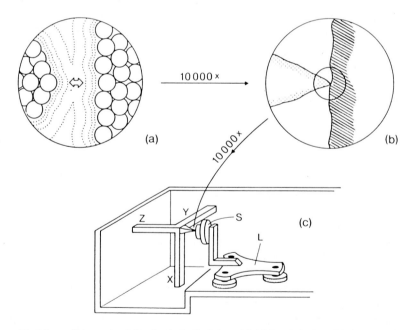

Fig. 1.1a–c. Schematic of the physical principle and initial technical realization of STM. (**a**) shows the apex of the tip (*left*) and the sample surface (*right*) at a magnification of about 10^8. The solid circles indicate atoms, the dotted lines electron density contours. The path of the tunnel current is given by the arrow. (**b**) Scaled down by a factor of 10^4. The tip (*left*) appears to touch the surface (*right*). (**c**) STM with rectangular piezo drive X, Y, Z of the tunnel tip at left and "louse" L (electrostatic "motor") for rough positioning (µm to cm range) of the sample S. From [1.34]

vertical resolution (perpendicular to the sample surface) of "only" 30 Å and a lateral (in-plane) resolution of "only" 4000 Å. High-resolution images in the tunneling regime with much smaller tip–surface separation were not obtained, mainly because of problems with vibration. Improved stability in metal–vacuum–metal tunneling experiments without scanning capability were later achieved by using a thermal drive apparatus [1.25], a differential-screw drive mechanism [1.26] or squeezable electron tunneling junctions [1.27–29].

The successful combination of vacuum tunneling with a piezoelectric drive system to a scanning tunneling microscope was first demonstrated in 1981 by *Binnig* et al. [1.30–34]. A conducting sample and a sharp metal tip, which acts as a local probe, were brought within a distance of a few ångstroms, resulting in a significant overlap of the electronic wavefunctions (Fig. 1.1). With an applied bias voltage (typically between 1 mV and 4 V), a tunneling current (typically between 0.1 nA and 10 nA) can flow from the occupied electronic states near the Fermi level of one electrode into the unoccupied states of the other electrode. By using a piezoelectric drive system for the tip and a feedback loop, a map of the surface topography can be obtained (Fig. 1.2). The exponential dependence of the tunneling current on the tip-to-sample spacing has proven to be the key for the high spatial resolution which can be achieved with the STM. Decreasing this spacing by only 1 Å typically leads to a one order of magnitude increase in the tunneling current. Under favourable conditions, a vertical resolution of hundredths of an ångstrom and a lateral resolution of about one ångstrom can be

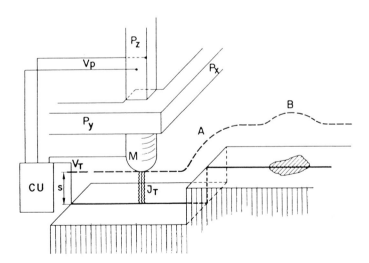

Fig. 1.2. Principle of operation of the STM. (Schematic: distances and sizes are not to scale.) The piezodrives P_X and P_Y scan the metal tip M over the surface. The control unit (CU) applies the appropriate voltage V_p to the piezodrive P_Z for constant tunnel current J_T at constant tunnel voltage V_T. The broken line indicates the z displacement in a scan over a surface step (A) and a chemical inhomogeneity (B). From [1.32]

reached. Therefore, STM can provide real-space images of surfaces of conducting materials down to the atomic scale. Its invention can be regarded as a milestone in surface science.

1.2 STM and Related Techniques

1.2.1 Local Proximal Probes

The scanning tunneling microscope (STM) has two main characteristic features: Firstly, it is a local probe which results from the particular geometry of the probing metal electrode being an extremely, i.e. atomically, sharp tip. Secondly, the STM is a proximal probe because of the close proximity of the probe tip and the sample surface required for obtaining the high spatial resolution.

Some local probes were already known before the invention of the STM. For instance, point contacts were used to locally measure superconducting energy gaps by tunneling experiments [1.35] or to study electron–phonon or electron–magnon interactions by the point contact spectroscopy technique [1.36, 37]. However, the fixed contacts prevented the collection of spatially resolved data. In contrast, the STM technique allows for probing surface properties at high spatial resolution since the probe tip, which is held at a few ångstroms from the sample surface, can be raster scanned by means of piezoelectric drives. Moreover, the tip can be positioned with atomic accuracy above a preselected surface site and a local experiment can be performed. This ability to perform local experiments together with the ability to characterize non-periodic surface structures can be regarded as the main advantages of STM compared with other surface analysis techniques. On the other hand, the accuracy of the determination of interatomic distances at periodic surfaces is much higher using surface diffraction techniques. The reason for this lies mainly with the piezoelectric drives used in STM showing hysteresis and creep, and in thermal drifts causing image distortions. Another problem with local probes is how representative the results obtained on a local scale are, or in other words when is it justified to speak of "typical" images or "typical" spectroscopic spectra. The complementary problem arises in the interpretation of results obtained by surface analysis techniques providing information averaged over macroscopic surface areas. In this case, the information obtained is averaged over many microscopic degrees of freedom and it is often unknown what the microscopic contributions are to what is measured macroscopically. Therefore, only a combination of local probes and averaging techniques can yield a complete picture.

The proximity of the probe tip to the sample surface in STM offers several advantages. Firstly, an electron optical lens system is not required compared to conventional electron microscopy, eliminating the problem of lens aberrations. Secondly, operation of a microscope in the near-field regime leads to a spatial resolution which is no longer wavelength limited. On the other hand, one has always to keep in mind that the close proximity of the probe tip to the sample surface can lead to modifications of what is intended to be measured. These modifications may either be due to the interaction between the probe tip and the sample surface, which is itself studied by a certain proximal probe, or by another interaction mechanism which is not intended to be probed. For instance, in STM the close proximity of the tip to the sample surface can lead to a modification of the local surface electronic structure. Particularly at small tip-to-surface distances, the perturbation of the electronic structure is significant and tip-induced localized states may be formed [1.38, 39]. Also, other tip–surface interactions may become relevant, e.g. the force interaction. This was first realized in STM experiments on relatively soft samples such as graphite [1.40]. The influence of forces may also show up in measurements of pressure-dependent sample properties [1.41, 42].

All the different types of possible interactions between a sharp tip in close proximity to a sample surface can in principle be used to invent novel STM-related scanning probe microscopies. For instance, the force interaction is used in the scanning force microscope [1.43]. The forces may have various origins (e.g. van der Waals forces, electrostatic forces, magnetic forces or frictional forces) yielding a further diversification of scanning probe microscopies. Other types of interactions may also be used, such as the thermal interaction leading to the scanning thermal profiler [1.44]. The spatial resolution which can be obtained with each of these scanning probe microscopies is mainly determined by the distance dependence of the interaction used in a particular scanning probe microscope.

1.2.2 Modes of Operation

Local proximal probes can be operated in a number of different modes. The most commonly used mode of operation, first introduced for the STM by Binnig, Rohrer and coworkers, is the constant current mode. In this mode, a feedback loop system forces the tip via a piezoelectric driver to be always at such a distance to the sample surface that the tunneling current flowing between these two electrodes remains constant. By recording the voltage which has to be applied to the piezoelectric driver in order to keep the tunneling current constant, i.e. recording the height of the tip $z(x, y)$ as a function of position, a topographical image can be obtained. The constant current mode can be used for surfaces which are not necessarily flat on an atomic scale, e.g. stepped surfaces (Fig. 1.3). The topographic height of surface features can be obtained directly provided that the sensitivity of the piezoelectric driver element is known. A disadvantage of the constant current (or, in general, constant interaction) mode is the finite response time of the feedback loop which sets relatively low limits for the scan speed.

To increase the scan speed considerably, another mode of operation has been introduced, the constant height mode [1.46]. In this mode, the tip is rapidly

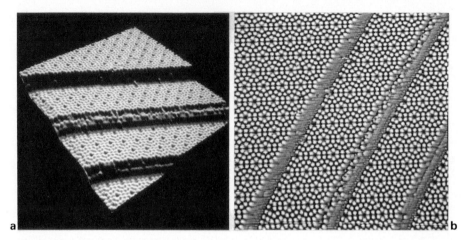

Fig. 1.3. (a) Perspective STM image of a ($320\,\text{Å} \times 360\,\text{Å}$) area on a Si(111)7 × 7 surface obtained in the constant current mode of operation. Three steps as high as four times a double-layer step separate narrow 7 × 7 reconstructed terraces. **(b)** Corresponding top-view image [1.45]

scanned at constant height over the sample surface while the feedback loop is slowed or turned off completely. The rapid variations in the tunneling current, which are recorded as a function of location, then contain the topographic information (Fig. 1.4). A significant advantage of this mode is the faster scan rate that can be reached because it is no longer limited by the response time of the feedback loop but only by the resonance frequencies of the STM unit. Consequently, image distortions due to thermal drifts and piezoelectric hysteresis can be reduced. Additionally, dynamic processes on surfaces can be studied better using this fast imaging mode. On the other hand, extracting the topographic height information from the recorded variations of the tunneling current in the constant height mode is difficult because the distance dependence of the tunneling current (or, in general, of any interaction) is often not known exactly. Another limitation of the constant height mode is that it is only applicable to atomically flat surfaces, otherwise the tip might crash into a surface protrusion while scanning at high speed.

Instead of scanning the tip over the sample surface, it is also possible to operate the STM in various tracking modes with the tip moving along selected paths [1.48]. For instance, profiles of steepest inclination, equal height or equipotential lines can be traced out.

Another mode of operation, differential microscopy, is based on a modulation technique [1.49]. The tip is made to vibrate in the scan direction (x direction) parallel to the sample surface at a frequency higher than the feedback response frequency. The differential image $dI(x, y)/dx$, which in the case of STM corresponds to the amplitude of the current modulation at the modulation frequency, often shows a significantly improved signal-to-noise

Fig. 1.4. Top-view STM image of a (70 Å × 70 Å) area on a C_8Cs-graphite intercalation compound surface obtained in the constant height mode of operation. Besides the atomic lattice of the graphite surface, a one-dimensional superlattice appears which is caused by the intercalated alkali metal [1.47]

ratio compared with the conventional constant current topography which can be recorded simultaneously.

Apart from these topographic modes of operation, the STM can also be used to get information about the spatially resolved local tunneling barrier height which is related to the work functions of both metal electrodes [1.50–52]. Therefore, the data obtained in this mode of operation have often been referred to as work function profiles. Again, the tip is vibrated at a frequency much higher than the feedback response frequency but now in the z-direction, which is perpendicular to the surface plane. Since the tunneling current I is exponentially dependent on the separation s between the tip and the sample surface: $I \propto \exp(-A\sqrt{\bar{\phi}}\,s)$ where $\bar{\phi}$ is the mean local tunneling barrier height and $A \approx 1$ Å^{-1} eV$^{-1/2}$, one gets $d\ln I/ds \sim -\sqrt{\bar{\phi}}$ if the distance dependence of the local tunneling barrier height is neglected. Therefore, by recording $d\ln I/ds(x, y)$, a map of the local tunneling barrier height can be obtained.

Further information from STM can be gathered by using spectroscopic modes of operation where the voltage dependence of the tunneling current is studied. The polarity of the applied bias voltage determines whether electrons tunnel into the unoccupied states of the sample (positive sample bias) or out of the occupied states (negative sample bias). The amount of the applied bias voltage determines which electronic states can contribute to the tunneling current. Various spectroscopic modes of operation have been introduced in the past, including modulation techniques in which the applied bias voltage is modulated at a frequency much higher than the feedback response frequency, or feedback interruption techniques in which the tip is fixed at a particular position (x, y, z) using a sample-and-hold circuit while the current versus voltage characteristic is measured. For a more detailed description of these spectroscopic modes of operation, the reader is referred to Chap. 5 [1.53, 54]. Here, it should only be emphasized that STM images may contain both atomic and electronic structure information, and to disentangle them is often not an easy task even by using additional spectroscopic modes of operation.

The various modulation and feedback interruption techniques can also be used in conjunction with STM-related scanning probe microscopies, although they are not yet as developed as for the STM.

1.3 Development of the Field

The field of STM and related scanning probe microscopies is still growing rapidly. This can best be illustrated by the statistics of STM conferences (Fig. 1.5). The rapid expansion of the field is mainly due to instrumental development and the application of these instruments to novel scientific disciplines. Instrumental development includes both the improvement of existing scanning probe microscopes (SPMs) and the invention of novel SPMs probing various interactions between a sharp tip and a sample surface on a local scale. Also, the availability of commercial instruments has significantly broadened the spectrum of users of STM and related SPMs. Initially, the application of STM

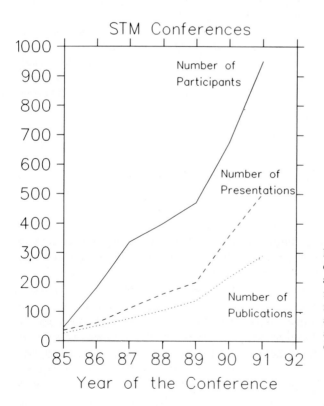

Fig. 1.5. Statistics of STM conferences between 1985 and 1991. Plotted are the number of participants, the number of presentations and the number of publications in the STM conference proceedings

was limited mainly to surface physics since only physicists knew how to build such instruments. In the meantime, SPMs have entered many other scientific disciplines such as chemistry, biology, metrology and materials science.

The development of the field of STM and related SPMs is also reflected in the "hot topics" at STM conferences. At the Oberlech (Austria) meeting in 1985 [1.55], the contributions mainly focussed on STM instrumentation. In 1986, at the first international STM conference in Santiago de Compostela (Spain) [1.56], everybody became excited about the refinements of the STM technique, particularly with regard to tunneling spectroscopy on semiconductor surfaces. Applications of STM in novel environments, such as in liquids or at low temperatures, and the development of atomic force microscopy were the hot topics of the 1987 conference held at Oxnard (USA) [1.57]. In 1988, during the conference in Oxford (GB) [1.58], it became clear that STM is also a powerful tool in surface chemistry and for studying molecules on surfaces. The applications of STM in biology and superconductivity (direct imaging of the Abrikosov flux lattice in a type II superconductor) dominated the 1989 conference in Oarai (Japan) [1.59]. In 1990, during the conference in Baltimore (USA) [1.60], variations of the STM technique based on the interaction of light with the tunneling junction or based on tunneling of spin-polarized electrons, and the manipulation of single atoms on surfaces with the STM, raised particular interest.

This enormous development within a few years has only been possible because of the fascination and excitement which STM and SPMs have raised in the scientific community. Saturation of this field is not yet in sight and one might ask: what comes next?

1.4 Prospects for the Future

Future progress in the field of STM and related SPMs may be expected in many different directions. However, there are some central problems which deserve special attention.

Firstly, a better knowledge and control of the microscopic structure of the probing tip is important in many respects, e.g. to understand measured corrugations in STM experiments [1.61, 62] or to control the influence of the tip's electronic structure on spectroscopic results [1.63–66]. Selective adsorption of atoms onto the tip may provide new possibilities in STM, particularly in view of the fact that the atom-resolved STM imaging process can be considered as a sequence of bond forming and bond rupturing [1.67]. Bringing the right atomic orbital in front of the tip may allow for atomic resolution studies in cases where other tips have failed to resolve atomic scale structures. The successful engineering of tips on an atomic scale was demonstrated some years ago using field ion microscopy (FIM) techniques [1.68, 69]. However, only a few combined

FIM–STM studies with direct observation of the tip's atomic structure before and after performing STM studies have been reported [1.70, 71]. The microscopic control of the tip's atomic structure is limited by the stability of the tip during the STM scanning process. Changes in the tip can, for instance, be deduced from discontinuities in STM images. However, the controlled engineering of tips on an atomic scale might also lead to more stable tips.

Another serious limitation of the STM technique so far is its lack of chemical sensitivity. As long as the STM is operated in the tunneling regime, the applied bias voltage has to be less than the surface work function which is of the order of 4 eV. This implies that only the valence states in an energy window of 4 eV around the Fermi level can be probed by STM. These valence states, however, take part in the chemical bond formation which itself can strongly modify the valence band structure. Therefore, valence band features are generally not specific for the elemental species in multi-component systems. This statement does not, however, exclude that in special cases where the direction of charge flow is well known, STM may be able to selectively image different atomic species, as shown for the GaAs(110) surface [1.72]. To attack the problem of the general lack of chemical sensitivity of STM, several solutions have been proposed, including the operation of STM in the field emission regime [1.73], performing vibrational spectroscopy at low temperatures [1.74] or using the interaction of STM with electromagnetic radiation [1.75]. However, chemical analysis on an atomic scale for any multi-component system by STM or related SPMs is yet to be shown.

The ability to manipulate single atoms by the STM has recently caused a lot of excitement [1.76]. It has become possible to build up artificial structures on surfaces, atom by atom. The experiment was performed in ultra-high vacuum and at low temperatures (4 K) but similar experiments may be possible in a more natural environment in the near future. The tip may, for instance, be used to bring two atoms or molecules close to each other and to image the reaction product, its shape and orientation, as a function of the chosen substrate. It may also be possible to build up atomic scale devices atom by atom by using an STM tip. These atomic scale devices may be based on the characteristics of atomic wave functions, as has recently been shown by a tunnel diode on the atomic scale [1.77–79]. A combination of such atomic scale devices with already existing microfabrication techniques may lead to a novel "nanoelectronics".

The prospects for the future of STM and related SPMs seems to be limited only by the imagination of the scientists themselves. The long journey into tunneling, which Esaki described in his 1973 Nobel Prize Lecture [1.10], is still continuing. Hopefully, this journey into tunneling may contribute to the benefit of mankind in a generalized sense which Esaki expressed in an impressive way: " . . . many high barriers exist in this world – barriers between nations, races and creeds. Unfortunately, some barriers are thick and strong. But I hope, with determination, we will find a way to tunnel through these barriers easily and freely, to bring the world together . . ." .

References

1.1   D.G. Walmsley: Surf. Sci. **181**, 1 (1987)
1.2   R. Oppenheimer: Phys. Rev. **31**, 66 (1928)
1.3   O.K. Rice: Phys. Rev. **34**, 1451 (1929)
1.4   R.H. Fowler, L. Nordheim: Proc. Roy. Soc. (London) A **119**, 173 (1928)
1.5   J. Frenkel: Phys. Rev. **36**, 1604 (1930)
1.6   R. Holm: J. Appl. Phys. **22**, 569 (1951)
1.7   J.G. Simmons: J. Appl. Phys. **34**, 1793 (1963)
1.8   J.G. Simmons: J. Appl. Phys. **34**, 2581 (1963)
1.9   L. Esaki: Phys. Rev. **109**, 603 (1958)
1.10 L. Esaki: Rev. Mod. Phys. **46**, 237 (1974)
1.11 I. Giaever: Phys. Rev. Lett. **5**, 147 (1960)
1.12 I. Giaever: Phys. Rev. Lett. **5**, 464 (1960)
1.13 J.C. Fisher, I. Giaever: J. Appl. Phys. **32**, 172 (1961)
1.14 I. Giaever, K. Megerle: Phys. Rev. **122**, 1101 (1961)
1.15 I. Giaever: Phys. Rev. Lett. **20**, 1286 (1968)
1.16 H.R. Zeller, I. Giaever: Phys. Rev. **181**, 789 (1969)
1.17 I. Giaever: Rev. Mod. Phys. **46**, 245 (1974)
1.18 B.D. Josephson: Phys. Lett. **1**, 251 (1962)
1.19 B.D. Josephson: Adv. Phys. **14**, 419 (1965)
1.20 B.D. Josephson: Rev. Mod. Phys. **46**, 251 (1974)
1.21 R. Young, J. Ward, F. Scire: Phys. Rev. Lett. **27**, 922 (1971)
1.22 R. Young, J. Ward, F. Scire: Rev. Sci. Instrum. **43**, 999 (1972)
1.23 R.D. Young: Phys. Today, Nov. 1971, p. 42
1.24 R.D. Young: Rev. Sci. Instrum. **37**, 275 (1966)
1.25 W.A. Thompson, S.F. Hanrahan: Rev. Sci. Instrum. **47**, 1303 (1976)
1.26 E.C. Teague: "Room Temperature Gold-Vacuum-Gold Tunneling Experiments"; Ph.D. Thesis, North Texas State University (1978), reprinted in: J. Res. NBS **91**, 171 (1986)
1.27 J. Moreland, S. Alexander, M. Cox, R. Sonnenfeld, P.K. Hansma: Appl. Phys. Lett. **43**, 387 (1983)
1.28 J. Moreland, P.K. Hansma: Rev. Sci. Instrum. **55**, 399 (1984)
1.29 P.K. Hansma: IBM J. Res. & Dev. **30**, 370 (1986)
1.30 G. Binnig, H. Rohrer, Ch. Gerber, E. Weibel: Appl. Phys. Lett. **40**, 178 (1982)
1.31 G. Binnig, H. Rohrer, Ch. Gerber, E. Weibel: Physica **109 & 110B**, 2075 (1982)
1.32 G. Binnig, H. Rohrer, Ch. Gerber, E. Weibel: Phys. Rev. Lett. **49**, 57 (1982)
1.33 G. Binnig, H. Rohrer: Helv. Phys. Acta **55**, 726 (1982)
1.34 G. Binnig, H. Rohrer: Physica **127B**, 37 (1984)
1.35 H.J. Levinstein, J.E. Kunzler: Phys. Lett. **20**, 581 (1966)
1.36 I.K. Yanson: Sov. Phys. JETP **39**, 506 (1974)
1.37 A.G.M. Jansen, A.P. van Gelder, P. Wyder: J. Phys. C**13**, 6073 (1980)
1.38 E. Tekman, S. Ciraci: Phys. Rev. B**40**, 10286 (1989)
1.39 S. Ciraci, A. Baratoff, I.P. Batra: Phys. Rev. B**41**, 2763 (1990)
1.40 J.M. Soler, A.M. Baro, N. Garcia, H. Rohrer: Phys. Rev. Lett. **57**, 444 (1986)
1.41 E. Meyer, R. Wiesendanger, D. Anselmetti, H.R. Hidber, H.-J. Güntherodt, F. Levy, H. Berger: J. Vac. Sci. Technol. A**8**, 495 (1990)
1.42 E. Meyer, D. Anselmetti, R. Wiesendanger, H.-J. Güntherodt, F. Levy, H. Berger: Europhys. Lett. **9**, 695 (1989)
1.43 G. Binnig, C.F. Quate, Ch. Gerber: Phys. Rev. Lett. **56**, 930 (1986)
1.44 C.C. Williams, H.K. Wickramasinghe: Appl. Phys. Lett. **49**, 1587 (1986)
1.45 R. Wiesendanger, G. Tarrach, D. Bürgler, H.-J. Güntherodt: Europhys. Lett. **12**, 57 (1990)
1.46 A. Bryant, D.P.E. Smith, C.F. Quate: Appl. Phys. Lett. **48**, 832 (1986)

1.47 D. Anselmetti, V. Geiser, D. Brodbeck, G. Overney, R. Wiesendanger, H.-J. Güntherodt: Synth. Met. **38**, 157 (1990)

1.48 D.W. Pohl, R. Möller: Rev. Sci. Instrum. **59**, 840 (1988)

1.49 D.W. Abraham, C.C. Williams, H.K. Wickramasinghe: Appl. Phys. Lett. **53**, 1503 (1988)

1.50 G. Binnig, H. Rohrer: Surf. Sci. **126**, 236 (1983)

1.51 G. Binnig, N. Garcia, H. Rohrer, J.M. Soler, F. Flores: Phys. Rev. B**30**, 4816 (1984)

1.52 R. Wiesendanger, L. Eng, H.R. Hidber, P. Oelhafen, L. Rosenthaler, U. Staufer, H.-J. Güntherodt: Surf. Sci. **189/190**, 24 (1987)

1.53 R.J. Hamers: Annu. Rev. Phys. Chem. **40**, 531 (1989)

1.54 R.M. Feenstra: "Scanning Tunneling Microscopy: Semiconductor Surfaces, Adsorption and Epitaxy", in *Scanning Tunneling Microscopy and Related Methods*, ed. by R.J. Behm, N. Garcia, H. Rohrer, NATO ASI Series E: Applied Sciences, Vol. 184 (Kluwer, Dordrecht 1990) p. 211

1.55 IBM J. Res. Dev. Vol. 30, pp. 353–572 (1986)

1.56 Proc. 1st Int. Conf. STM'86, ed. by N. Garcia, Surf. Sci. **181**, pp. 1–412 (1987)

1.57 Proc. 2nd Int. Conf. STM'87, ed. by R.M. Feenstra, J. Vac. Sci. Technol. A**6**, pp. 259–556 (1988)

1.58 Proc. 3rd Int. Conf. STM'88, ed. by W.M. Stobbs, J. Microsc. **152**, pp. 1–887 (1988)

1.59 Proc. 4th Int. Conf. STM'89, ed. by. T. Ichinokawa, J. Vac. Sci. Technol. A**8**, pp. 153–720 (1990)

1.60 Proc. 5th Int. Conf. STM'90/NANO-I, ed. by R.J. Colton, C.R.K. Marrian, J.A. Stroscio, J. Vac. Sci. Technol. B**9**, pp. 403–1407 (1991)

1.61 A. Baratoff: Physica **127B**, 143 (1984)

1.62 C.J. Chen, Phys. Rev. Lett. **65**, 448 (1990)

1.63 R.M. Tromp, E.J. van Loenen, J.E. Demuth, N.D. Lang: Phys. Rev. B**37**, 9042 (1988)

1.64 J.E. Demuth, U. Koehler, R.J. Hamers: J. Microsc. **152**, 299 (1988)

1.65 Sang-il Park, J. Nogami, H.A. Mizes, C.F. Quate: Phys. Rev. B**38**, 4269 (1988)

1.66 T. Klitsner, R.S. Becker, J.S. Vickers: Phys. Rev. B**41**, 3837 (1990)

1.67 C.J. Chen: J. Phys.: Condens. Matter **3**, 1227 (1991)

1.68 H.-W. Fink: J. Res. & Dev. **30**, 460 (1986)

1.69 H.-W. Fink: Phys. Scr. **38**, 260 (1988)

1.70 Y. Kuk, P.J. Silverman, Appl. Phys. Lett. **48**, 1597 (1986)

1.71 T. Sakurai, T. Hashizume, I. Kamiya, Y. Hasegawa, T. Ide, M. Miyao, I. Sumita, A. Sakai, S. Hyodo: J. Vac. Sci. Technol. A**7**, 1684 (1989)

1.72 R.M. Feenstra, J.A. Stroscio, J. Tersoff, A.P. Fein: Phys. Rev. Lett. **58**, 1192 (1987)

1.73 B. Reihl, J.K. Gimzewski: Surf. Sci. **189/190**, 36 (1987)

1.74 D.P.E. Smith, M.D. Kirk, C.F. Quate: J. Chem. Phys. **86**, 6034 (1987)

1.75 L.L. Kazmerski: paper presented at the 5th Int. Conf. STM'90/NANO-I, Baltimore, USA (1990)

1.76 D.M. Eigler, E.K. Schweizer: Nature **344**, 524 (1990)

1.77 P. Bedrossian, D.M. Chen, K. Mortensen, J.A. Golovchenko: Nature **342**, 258 (1989)

1.78 I.-W. Lyo, Ph. Avouris: Science **245**, 1370 (1989)

1.79 Ph. Avouris, I.-W. Lyo, F. Bozso, E. Kaxiras: J. Vac. Sci. Technol. A**8**, 3405 (1990)

2. The Rise of Local Probe Methods

H. Rohrer

Scanning tunneling microscopy (STM) instigated the rapid development of an expanding family of local probe microscopies or, more appropriately, of scanning probe methods. Their basic ingredient is a local experiment which, repeated at sequential locations, can be assembled to form an image. This is also how STM developed. The original goal was to learn about the local structural, electronic, and growth properties of very thin insulating layers, in particular at tunnel junctions. "Local" meant on the scale of the inhomogeneities of those properties, which were believed to be no larger than a few nanometers in size, a scale that was entirely inaccessible with existing techniques. Electron tunneling appeared to be a promising approach, provided it could be done locally. This led in a natural, non-premeditated way to the local probe method "scanning tunneling microscopy". Electron tunneling already contained two of the four major technical elements of a local probe method: a strongly distance-dependent interaction and, inherently necessary, close proximity of probe and object. One tunneling electrode in the form of a sharp conducting tip would provide the third element, the local probe. Metal tips with a radius of curvature of about 20 nm, which would have brought the resolution to the desired level, were already in use as field emitters and in field ion microscopy. These three elements determine the resolution. The fourth element, finally, was the stable positioning of the probe with respect to the object with an accuracy better than the desired resolution and within the practical range of the interaction. We expected to achieve this with piezo drives made from commercially available material.

Although the development of STM appears straightforward in this condensed retrospective view, it nevertheless required some ideas and effort; we had mistakes to correct and, in particular, had to deal with many unknowns. For instance, replacing the well-defined field emission tip with a simple ground-metal tip simplified matters and this could – and did – help us achieve atomic resolution since, unless specially prepared, most tips end up with one atom, thus becoming an atomic size probe. However, it was by no means clear that the apex of such a tip would be mechanically stable. Indeed, it was usually unstable in the beginning of STM; nowadays there are nearly as many recipes for obtaining stable tips as there are scientists using them. The same goes for the piezo drives. With atomic resolution in sight, the tip position had to be controlled within a fraction of an ångstrom, not just of a nanometer. Only the experiments themselves showed afterwards that the response of the piezo to an applied voltage was continuous at least down to the picometer level. It is remarkable how often success is the reward for trying the unknown.

At the beginning, the anticipated resolution only matched that of scanning electron microscopy as far as the structural properties were concerned. But although STM did not provide better resolution, it offered something else. A local tunneling experiment, e.g. tunneling spectroscopy, contains a wealth of information and reflects local electronic and chemical properties, in addition to the structural ones. This and the conceptual simplicity of the approach were sufficient reason to start, and by the time we had finished our first successful experiment, the resolution was nearly at an atomic level. This is reminiscent of the story of electron microscopy, which at the beginning offered poorer resolution than optical microscopy, was more complicated and even destroyed most of the samples in the imaging process. Could we imagine present-day science and technology without electron microscopy? We might claim that what is different constitutes progress, and not so much what is "better".

STM was not developed from one of the already existing local probe methods or from ideas about them, nor was it done in the community of microscopists or in other circles with the appropriate competence. No technically new component or new material was necessary, no new physical insight was required and no additional theoretical basis had to be established, yet somehow the belief prevailed in these communities that "it" could not be done. Vacuum tunneling apparently crossed many a mind but was dismissed as unfeasible. The topografiner came the closest. Stylus profilometry did not go beyond carefully shaped, smoothed, and well-defined sensing tips of a radius of curvature of about a micrometer and, therefore, stayed in the micrometer resolution range. Instead, a splinter of diamond accompanied by a few ideas brought atomic force microscopy with atomic resolution. We heard so many objections, for example to the positioning of a local probe with subångstrom accuracy, including objections citing the uncertainty principle, even after the STM had worked! We might learn from this that an occasional change in the field of interest can bring unexpected progress.

What initially appeared rather exotic with a competitive component is now considered for what it is, namely a new method with an ever expanding variety of new, exciting possibilities. Local probe methods are now generally accepted as a central, stimulating approach to science and technology on the nanometer scale. The particularly appealing aspects of local probe methods are their conceptual simplicity, the variety of probing interactions and thus the local properties accessible, and the range of applicability to metals and insulators, in ultra-high vacuum to electrolytes, and at high temperatures to cryogenic liquids. The first ten years were merely a beginning. Observation and the understanding of local properties, the essence of microscopy, are but a starting point. We can already see the development going from an analytical method to an active tool to manipulate and modify individual nanometer-size objects and functional units, and to control their functions, to interface so to speak the macroscopic with the nanoscopic world. Such a richness could not possibly be foreseen when setting out to build an STM and not even after the first STM images showed atomic-size features. It is, therefore, moot to speculate on future developments, but they are sure to be as exciting as those of the past ten years. The devoted endeavors of many a scientist will be crucial.

3. STM on Metals

Y. Kuk

With 19 Figures

This chapter presents a review of studies of clean metal surfaces by scanning tunneling microscopy and spectroscopy. The operating principles of the scanning tunneling microscope for small metallic corrugations are explained. Various spectroscopies are described and compared with theory. Some examples of past accomplishments on metal surfaces are given.

While some metal surfaces undergo reconstruction in order to lower their surface energy, most reveal bulk-like terminations at room temperature [3.1]. Direct imaging of geometric and electronic surface structures with atomic resolution had been a dream before the invention of scanning tunneling microscopy by *Binning, Rohrer* and co-workers [3.2]. Since the first report of Si(111)–(7 × 7) [3.3], a structure which had been an unsettled question for more than 20 years, scanning tunneling microscopy has been accepted as a powerful surface science tool. Soon after the Si(111)–(7 × 7) image, the reconstructed Au(110)–(1 × 2), another well-studied surface, was imaged by the inventors' group [3.4], confirming the previously proposed structure. Structures of clean and chemisorbed metal surfaces have been reported by many groups since then. In this chapter, scanning tunneling microscopy and spectroscopy on clean metal surfaces will be discussed; structures of adsorbate covered metal surfaces will be dealt with in Chap. 4.

In scanning tunneling microscopy, the three-dimensional variation of charge density at a surface is probed via electron tunneling between a sharp tip and the sample. This vacuum phenomenon, electron tunneling, has been known since the introduction of quantum mechanics in the 1920s [3.5, 6]. When the tunneling gap is small and the voltage low, the relation of the tunneling current to the gap distance can be simplified to

$$I \propto (V/s) \exp(-A\bar{\phi}^{1/2} s), \tag{3.1}$$

where $A = 1.025 \, (\text{eV})^{-1/2} \, \text{Å}^{-1}$, $\bar{\phi}$ is the average barrier height between the two electrodes, V is the bias potential between the sample and the tip, and s is the gap distance. Equation (3.1) indicates that a 1-Å change in the gap distance produces roughly one order of magnitude change of the tunneling current with $\bar{\phi} \sim 4 \, \text{eV}$. This exponential dependence was first measured by *Young* et al. [3.6] and *Teague* [4.7]. Later, *Binnig* et al. [3.2] realized that surface structures could be mapped by using a feedback system to maintain a constant tunneling current under ultra-high vacuum (UHV) conditions, resulting in unprecedented atomic resolution.

While large charge density corrugations ($\sim 1\,\text{Å}$) are typically observed on semiconductor surfaces due to the presence of dangling bonds, those on metal surfaces are small ($< 0.1\,\text{Å}$), as measured by helium diffraction experiments, unless they are reconstructed or chemisorbed [3.8]. Atomic imaging of metal surfaces, therefore, requires high lateral and vertical resolution [3.9]. Unusually large corrugations have been observed on the close-packed metal surfaces with a small tunneling gap; the details and the possible mechanism for this will be discussed in Sect. 3.3. In this chapter, the role of the tunneling tip and its particular importance in metal studies will first be discussed in Sect. 3.1. The principle of scanning tunneling spectroscopy (STS) and some examples will follow. Finally, a few examples will be presented to demonstrate the achievements of STM on metal surfaces.

3.1 Tunneling Tip

It has been known that a one-atom tip is required to obtain a well-resolved STM image of a bulk-terminated metal surface. However, the role played by the tunneling tip is still not clear. The size, shape and chemical identity of the tip influence not only the resolution and shape of an STM scan but also the measured electronic structure. Experimentally, tips have been prepared by mechanical grinding or chemical etching from a variety of materials, most often W. Since a bcc crystal has a lower surface energy on the (110) faces, $\langle 100 \rangle$ and $\langle 111 \rangle$ oriented tips (single crystal wires) can be sharpened by annealing in the presence of a high electric field [3.10–12], producing a pyramid shape of (110) facets. When tip annealing and high electric field are not available, a high tunneling current (100 nA–10 µA) is known to often improve the characteristics of a tunneling tip. These techniques are supposed to produce a sharp, clean, and symmetric tip, but asymmetric or double tips [3.13] are often formed, resulting in misleading sample topographs.

One experiment has been reported [3.10] in which a field ion microscope (FIM) was used to examine the geometry of a tunneling tip, although other combinations of STM and FIM have since been assembled [3.14, 15]. By taking FIM images of the tunneling tip before and after an STM scan, the character of the STM topograph can be correlated to the tip structure. The dependence of the measured corrugation amplitude on the size of the tunneling tip has been experimentally examined. Figure 3.1a shows an FIM picture of the tunneling tip used in the STM scan of the clean Au(001)–(5 × 20) surface (this surface structure will be discussed later) with corrugation width of 14.4 Å [Fig. 3.1b]. The corrugation amplitude was estimated theoretically [3.16] as

$$\Delta \propto \exp\left[-\beta(R + s)\right], \tag{3.2}$$

where $\beta \simeq (1/4)\kappa^{-1} G^2$. R is the tip radius of curvature (approximated as a hemisphere), s is the gap distance, κ^{-1} is the electron decay length in

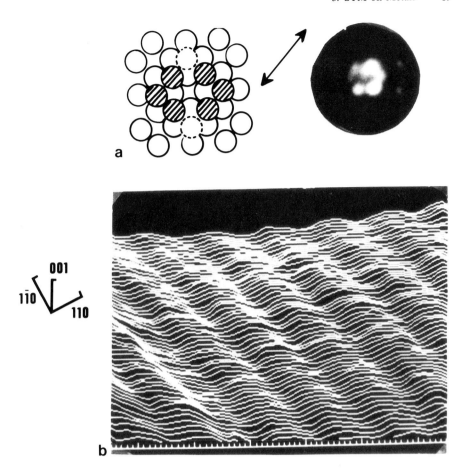

Fig. 3.1. (**a**) FIM image of the W(100) tip before and after the STM scan. The arrow indicates the scanning direction with respect to the tip. Filled circles are the first layer atoms. (**b**) STM topograph of Au(001). Rows along the close-packed $\langle 1\bar{1}0 \rangle$ are separated by ~ 14 Å

the vacuum, and G is the smallest surface reciprocal wave vector ($2\pi/a_i$, and a_i is the largest corrugation width). Average measured corrugations of the Au(100)–(5 × 20) and the Au(110)–(1 × 2) reconstruction [3.17] as a function of the tip size are summarized in Fig. 3.2. The measured β for the Au(100) and Au(110) are in good agreement with the values predicted by (3.2). Tip size vs. measured corrugation for several other surfaces is shown. Broken horizontal lines in Fig. 3.2 indicate STM noise levels, due to mechanical vibration or electronic noise, for two hypothetical STMs. The significance of the tip size is now apparent from Fig. 3.2; the tip must be terminated by a single atom in order to detect any reconstruction with the noisier STM. On the other hand, the corrugations of the Au(100), Au(110), and Si(111)–(7 × 7) surfaces could be detected by a 20-Å tip, with an STM noise level of 0.05 Å. In order to image most

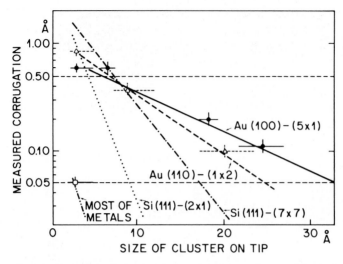

Fig. 3.2. Dependence of the measured corrugation on the size of the tip for Au(001)–(5 × 1) (solid line and filled circle), Au(110)–(1 × 2) (broken line and triangle), Si(111)–(7 × 7) (—·—), Si(111)–(2 × 1) (dotted line), and most (1 × 1) metals (—··—)

metals with small reconstruction unit cells or (1 × 1) surfaces, a very sharp tip and low STM noise level are required.

3.2 Tunneling Spectroscopies

Many metal surfaces show smaller corrugations than semiconductor surfaces by helium diffraction, and those measured by scanning tunneling microscopy are even smaller, since the tunneling tip is farther from the surface than the turning point of the He atoms. Because the decay length varies with electron state (much longer for an s electron than a d electron), an STM image of a metal surface is mainly due to its s-state electrons. On semiconductor surfaces, on the other hand, surface charge density is dominated by dangling bond states (p states), showing large corrugation. In this section, tunneling measurements of local barrier height and electronic density of states will be described to emphasize the difference between metal and semiconductor surfaces.

3.2.1 Current Versus Gap Distance

One direct indication of metal–vacuum–metal tunneling is the exponential dependence of the tunneling current as the tunneling gap increases, as described by (3.1) and first reported by *Young* et al. [3.6]. Stability obtained by good vibration isolation is thus crucial in the tunneling junction. The continuous exponential variation over four orders of magnitude has been demonstrated by *Binnig* et al. [3.2]. They reported an average barrier height of ∼ 3.2 eV for the junction between a W tip and a Pt plate, demonstrating an impurity-free junction.

The *I–s* relation on a reconstructed surface was studied by *Kuk* and *Silverman* [3.18]. Figure 3.3 shows *I–s* and *ϕ–s* spectra on a Au(100)–(5 × 20) surface. As reported earlier by *Gimzewski* et al. on a Ag film [3.19] and *Dürig* on a Ir film [3.20], a near exponential dependence of *I* as a function of *s* is shown. The contact point (*x* = 0 in Fig. 3.3) was defined at the jump in the tunneling current. From gap distance dependence of this slope, the local barrier height can be obtained from

$$\bar{\phi} = 0.952 \, (d \ln I/ds)^2 \,, \tag{3.3}$$

where *s* is in Å. The result is in good agreement with the theoretical work by *Lang* [3.21] for a one-atom Na tip, where local barrier height was calculated as a function of gap distance. With *s* > 5 Å, the barrier between the tip and the sample can be regarded as a small perturbation of two independent metallic states, but for *s* < 5 Å the perturbation is substantial. From the contact of the sample and the tip, a contact resistance was estimated to be ∼ 24 kΩ in good

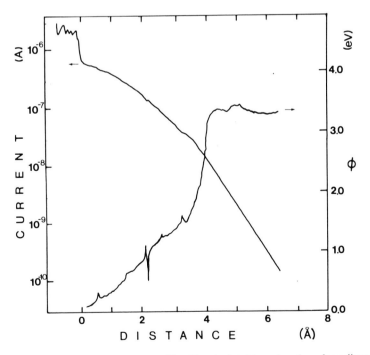

Fig. 3.3. The tunneling current and local barrier height as a function of gap distance in semilog plot. A contact point is defined as 0. The tip voltage is 50 mV

agreement with theoretical prediction [3.21, 22]. The local barrier height and the contact resistance were found to be a strong function of the shape and size of the tip, as estimated from the topography after contact was made.

3.2.2 Electronic Structure by *dI/dV*

The relation between the tunneling current and bias voltage for metals can be simplified by treating both electrodes as free electron-like metals. The vacuum tunneling between the tip and sample can then be divided into two regimes. When the bias voltage is smaller than the work function, the tunneling current is represented by (3.1). The current is proportional to bias voltage and exponentially dependent on the gap distance. When $V > \bar{\phi}$, the tunneling current dependence on the bias voltage is described by the Fowler–Nordheim relation [3.5]. The tunneling current in this field emission regime experiences the positive kinetic energy region of the vacuum gap. Depending on the gap distance and bias voltage, the tunneling electrons form a standing wave in this region, resulting in an oscillatory transmission probability. Figure 3.4 shows the oscillations in *dI/dV* from 5 to 18 V with a constant tunneling current of 1 nA. Recently, this field emission resonance has been studied in detail by *Coombs* and *Gimzewski* [3.23] and *Kubby* [3.24] to enhance some topographic features and understand light emission from the tunneling junction.

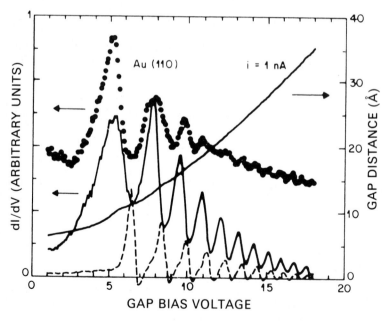

Fig. 3.4. *dI/dV* (closed circles: experimental; solid curve: theoretical) and gap distance vs. bias voltage at the tunneling current of 1 nA. From [3.59]

While a normal STM image is a convolution of geometric and electronic information about the surface, the electronic information can also be separately measured [3.25, 26]. With the tunneling tip poised over a region of interest and the gap fixed by opening the feedback loop momentarily, the bias voltage can be ramped to measure the tunneling current as a function of applied voltage. This measurement can also be performed at each point of a topography scan, resulting in spatially resolved $I(V)$ relations. Interpretation of these spectra is quite similar to metal–insulator–metal tunneling, replacing the insulator with a finite vacuum gap. For small bias voltage ($\bar{\phi} > V$), the tunneling current can be written as

$$I \propto \int_0^{eV} \varrho(E)D(E, V)dE \ , \tag{3.4}$$

where $\varrho(E)$ is the sample surface density of states and $D(E, V)$ is the transmission coefficient of the barrier at voltage V. D can be calculated by the WKB method for free electron model; the result is shown in Fig. 3.5. The transmission coefficient can be approximated by

$$D(V) = \alpha V + \gamma V^3 + \cdots \ , \tag{3.5}$$

where the cubic correction term is more apparent at smaller gap distances in Fig. 3.5. The local density of states can be deduced from dI/dV (differential conductance) in the low voltage limit by

$$dI/dV \propto \varrho(r, V)D(V) \ , \tag{3.6}$$

where $\varrho(r, V)$ is the local density of states of the sample evaluated at the center of the tip (r). The relation dI/dV vs. V has been shown to be proportional to the

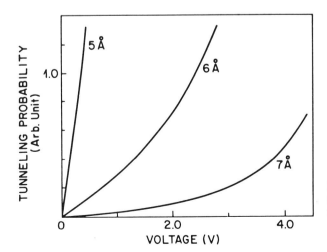

Fig. 3.5. $D(V)$–V at 3 different gap distances calculated by the WKB method

density of states in the low voltage limit in metal–insulator–metal tunneling [3.27]. A plot of $d\ln I/d\ln V$ removes the parabolic dependence introduced by the cubic term in (3.5), so the peak in this curve corresponds roughly to resonances in the tip and sample densities of states, although their exact heights and positions also depend on the gap distance [3.28, 29]. When the tunneling gap is small ($< 5\,\text{Å}$), the perturbation is too large to use (3.6), which is based on the transfer Hamiltonian where the bases are two independent solutions of the tip and sample Hamiltonian. Unusually large corrugations observed on several close-packed metal surfaces cannot be explained by this theory.

A series of $d\ln I/d\ln V$–V curves measured on Au(100)–(5 × 20) at various gap distances are plotted in Fig. 3.6 [3.18]. The gap distances were estimated from the I–s relation taken after the I–V. The scanning tunneling spectra at different gaps show a resemblance but the peak widths and heights vary with the gap distances. An earlier photoemission measurement for this surface [3.30] showed a peak near $\sim -0.7\,\text{eV}$, near the Brillouin zone boundary. A slight downward shift of the tunneling peak positions vs. the photoemission spectroscopy data has been observed by several groups, but the reason is not fully understood. The $-0.7\,\text{eV}$ peak has been ascribed to a bulk sp band. Although there are other peaks with higher intensity from d bands, there is little evidence in Fig. 3.6 of d-band contribution, which would be strongly dependent on gap distance since the electron decay length is much shorter than that in the sp band. Comparison of STS to (inverse) photoemission demands special care, since the surface states, overwhelmed by bulk states in (inverse) photoemission, can still

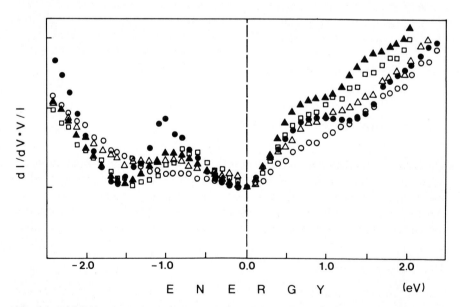

Fig. 3.6. $V/I * dI/dV$–V spectra at various gap distances: 6.0 (filled circle), 6.6 (filled triangle), 7.5 (open square), 8.1 (open triangle), and 8.6 Å (open circle)

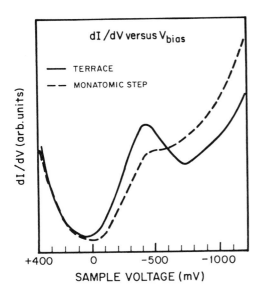

Fig. 3.7. dI/dV on terraces of the Au(111) and near a step edge. From [3.63]

be probed by tunneling spectroscopy. In addition, STS measures the local density of states summed over k vectors with varying weighting factors due to different decay lengths at various k in the presence of the tunneling tip [3.16]. Tunneling spectroscopy of the Au(111) surface shows a state at ∼ 0.4 eV above the Fermi level. The peak height and width of the surface state vary spatially. For example, the peak is smaller near the step edge as shown in Fig. 3.7. By imaging near the surface state, better contrast of the Au(111) stacking fault reconstruction could be obtained [3.31]. As these two examples demonstrate, most surface states on metals are not as sharp or high as those on semiconductors, which are mainly due to directional dangling bond states.

3.3 Examples on Metal Surfaces

The scanning tunneling microscope was initially used to simply image surfaces. As tunneling spectroscopy became widely used, STS data has been employed to refine surface structural models derived from scanning tunneling microscopy alone. By adding adsorbate sources (gas or solid phase) and temperature control, the study of dynamical behavior on metal surfaces has become possible.

3.3.1 Surface Structures

One of the most notable achievements of scanning tunneling microscopy has been the elucidation of the atomic arrangements of various surfaces. Many surface structures, which had been disputed by various experimental techniques and theoretical calculations, have been resolved by the direct imaging capability of the scanning tunneling microscopy. The first STM study of a metallic surface was done on Au(110) [3.2]. The (110) surfaces of Au, Pt, and Ir have been shown by many surface science techniques [3.32–35] to exhibit missing row-type reconstruction resulting in (1×2) structures. STM studies confirmed the missing row structure [3.4, 36] and examined the order–disorder phase transition [3.37]. The topograph of Au(110) (Fig. 3.8) shows the (1×2) reconstruction with alternate $\langle 1\bar{1}0 \rangle$ missing rows. As the sample temperature increases, the (1×2) structure had previously been observed to undergo a phase transition to a bulk-like (1×1) in the diffraction pattern [3.38]. Theoretical and LEED studies [3.38, 39] have indicated that this is an order–disorder transition of a 2-D Ising universality class. In the disordered (1×1) phase, the top layer atoms are

Fig. 3.8. $190 \times 240\text{-Å}^2$ gray scale topograph of the Au(110)–(1 × 2) surface after annealing at 600 K

commensurate with bulk lattice sites, i.e. the surface is disordered in the lattice–gas sense [3.38]. By diffraction methods, the transition temperature for the Au(110) was found to be ~ 700 K and is very sensitive to surface impurities. Figure 3.9 shows a topograph of the quenched Au(110) surface after annealing within 10 K of the phase transition temperature, as determined by the LEED patterns. At this temperature, although the STM topograph shows that the (1×2) reconstruction with small domain sizes is still present, the half order spots have almost disappeared in the corresponding LEED pattern. This is not surprising since the average domain size is 20–40 Å, far less than the usual coherence length for electron diffraction. The domains are separated not only by steps but also by (1×3) or (1×4)-type missing row structures. The corresponding diffraction pattern of the observed STM topograph can be calculated by a Fourier transformation. Figure 3.10 shows the calculated diffraction intensities just below the phase transition temperature. Since the (1×2) reconstructions in adjacent terraces separated by a monatomic step have a $\pi/2$ phase shift, interference of adjacent terraces results in the displacement of the half order peak from the normal position and also splits the $(1, 0)$ peak. A similar result was observed by X-ray diffraction and explained by the presence of steps [3.33].

The close-packed plane of fcc metal surfaces such as Au(111) [3.40, 41] and Al(111) [3.42] have been imaged with atomic resolution (Fig. 3.11). Unreconstructed (1×1) structures of other surfaces have been reported [3.43–45]. These unusually high spatial resolutions and large corrugations cannot be explained by the presently accepted transfer Hamiltonian approach [3.16]. Several mechanisms have been proposed: 1) highly localized surface states present near the Fermi level, 2) presence of an atom with an unusual electronic state (for example an atom with a p_z state) on the apex of the tunneling tip [3.46]

Fig. 3.9. 225×100-Å^2 gray scale topograph of the Au(110)–(1×2) surface after annealing at 700 K

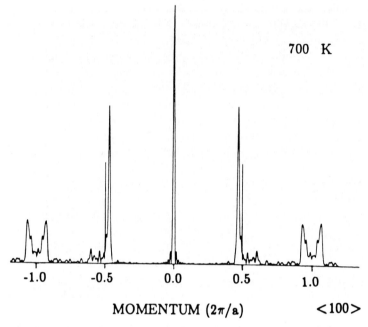

700 K

MOMENTUM (2π/a)   **<100>**

Fig. 3.10. Diffraction intensity along the ⟨100⟩, calculated from the STM image of Fig. 3.9

and 3) influence of atomic forces between the tip and sample. However, these large corrugations may well be explained by a new tunneling theory which includes a strong perturbation. Figure 3.12 shows a topograph of the Au(111) surface with atomic resolution. This surface also exhibits a long-range reconstruction of $(22-23 \times \sqrt{3})$, which is caused by stacking faults between fcc and hcp, and shows a surface state $\sim 0.4\,\text{eV}$ above the Fermi level. The higher atomic density and change of stacking sequence was confirmed by the deposition of Ni on the surface [3.47].

The (100) surfaces of Au and Pt have been known to reveal "(5 × 20)" and "(5 × 12)" reconstructions, respectively. These reconstructions are caused by a close-packed triangular first layer with a higher atomic density than the bulk. An earlier study by *Binnig* et al. [3.48] proposed "(26 × 68)" based on the length calibration of the surface, but recent topography by the author's group (Fig. 3.13) [3.36] shows an atomic image of the surface, suggesting four domain structures represented by

$$\begin{bmatrix} 5 & 0 \\ 1 & 20 \end{bmatrix}, \begin{bmatrix} 5 & 1 \\ 0 & 20 \end{bmatrix}, \begin{bmatrix} 20 & 0 \\ 1 & 5 \end{bmatrix}, \begin{bmatrix} 20 & 1 \\ 0 & 5 \end{bmatrix}.$$

The domains are separated by regularly arranged misfit dislocations and slight buckling of the atomic rows caused by a high atomic density in the first layer. Details of the Pt(100) surface (5 × 12) reconstruction are similar, as reported by *Behm* et al. [3.49].

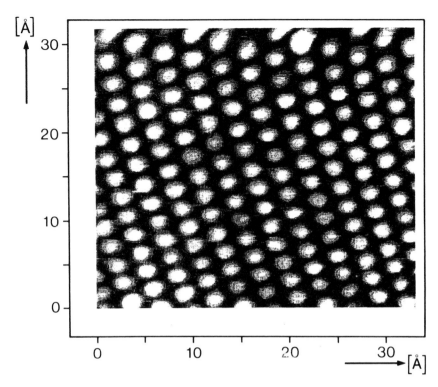

Fig. 3.11. 34×34-Å^2 gray scale topograph of the clean Al(111) surface. From [3.60]

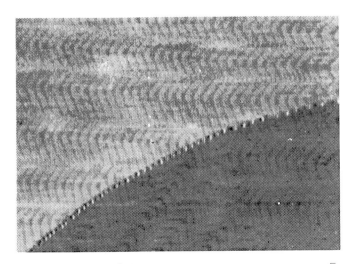

Fig. 3.12. 3000×2000-Å^2 gray scale topograph of the Au(111)–$(22 \times \sqrt{3})$, shown as a herringbone pattern. From [3.61]

Fig. 3.13. 160×40-Å^2 gray scale topograph of the Au(100) surface

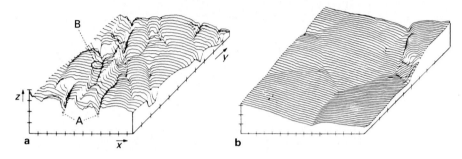

Fig. 3.14. 600×550-Å^2 topograph of Ag films condensed at (**a**) 80 K and (**b**) 300 K. From [3.62]

Atomic resolution is not required, or even desirable, in all scanning tunneling microscopy studies. There are many important and interesting applications for scanning tunneling microscopy with nanometer resolution on practical surfaces. Roughness measurements of metallic layers have been studied by some groups [3.50, 51]. For example, silver films condensed at room temperature and 90 K exhibit a difference in roughness (Fig. 3.14) which can be explained by diffusion limited aggregation.

3.3.2 Dynamics

Phase transitions of metal surfaces can be induced by changing the sample temperature and introducing adsorbates. Surface phase transitions have received increasing attention as equilibrium surface structures are better understood. Structural, order–disorder, magnetic, melting, and roughening transitions have been studied by a variety of surface sensitive techniques [3.52]. Other dynamic phenomena, such as surface diffusion and epitaxial growth, can be understood in light of these transitions. In order to study the dynamics of these phenomena by scanning tunneling microscopy, the scanning time must be shorter than the characteristic time constants of these transitions and the STM should be operable at elevated and cryogenic temperatures. When these conditions are not met, equilibrium structures may be studied by quenching samples from elevated temperatures. At present, studies of structural phase transitions, surface diffusion, adsorbate-induced transitions, and epitaxial growth have been reported.

Surface diffusion has been studied by FIM to obtain activation energies between atomic sites and by marker or tracer techniques to deduce macroscopic diffusion coefficients. Since STM can image a wider viewing area than FIM, it can be useful in the study of surface diffusion. The diffusion of step edges on Au(111) surfaces was observed in UHV and in air [3.53]. A series of time-lapse STM topographs of indentations or protrusions created by the tunneling tip show some movement of step edges in Fig. 3.15. With known markers (a circle or a line), the surface diffusion coefficient can be calculated from the diffusion equation. The hopping motion of individual atoms is faster than the STM scanning speed, so only the scratch-decay method [3.53, 54] can be used at room temperature on most metals. At low temperature, it is possible to estimate the local surface diffusion coefficient and activation barriers by tracking individual atoms.

Similarly, close observation of step edges by scanning tunneling microscopy at elevated temperature permits the study of surface roughening transitions [3.55]. Above a critical temperature, roughening appears as step meandering and step height variation. In many diffraction techniques, determination of the roughening temperature and even proof of its existence is very difficult because

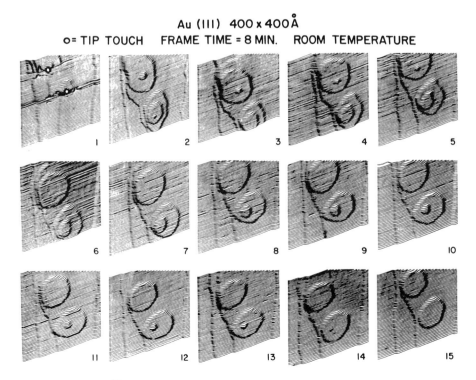

Au (III) 400 x 400 Å
o = TIP TOUCH   FRAME TIME = 8 MIN.   ROOM TEMPERATURE

Fig. 3.15. 400×400-Å^2 topographic images of Au(111) surface taken every 8 minutes. The indentation was made at frame 1 by the tunneling tip. From [3.63]

of the smooth change of the line shapes. In scanning tunneling microscopy, however, the topography of the step edges can be imaged directly. Figure 3.16 shows gray scale STM images of the Ag(115) surface at 20, 58, 98, and 145 °C. In images at 20 and 58 °C, there are large(115) terraces with low step densities. Above 98 °C all steps have a large number of thermally generated kinks. The

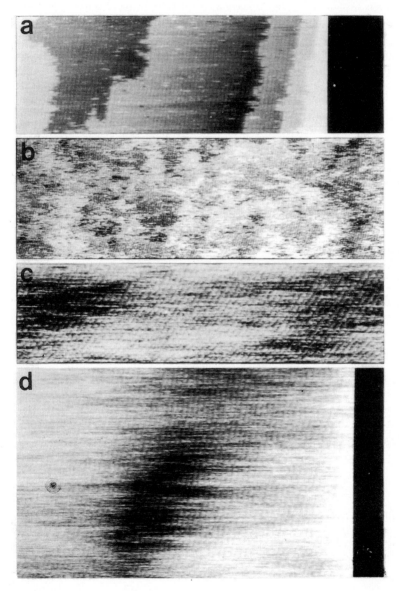

Fig. 3.16. Gray scale images of the Ag(115) surfaces taken at (**a**) 20°C (585 × 220-Å²), (**b**) 58°C (1180 × 375-Å²), (**c**) 98°C (450 × 120-Å²), and (**d**) 145°C (465 × 280-Å²). From [3.64]

meandering of steps clearly indicates that the Ag(115) surface is already in the roughened state. Although this study only shows snap shots at various temperatures, it demonstrates the feasibility of the study of phase transitions by scanning tunneling microscopy.

Initial stages of epitaxial growth on metals and semiconductors have been widely studied recently. STM experiments can yield not only the structural information of overlayer films but also the dynamics of growth when the overlayer diffusion is sufficiently slow. While the details of the Au overlayer on Ni(110) could not have been determined by other surface science techniques, the (7×4) structure with a $c(2 \times 4)$ subunit structure was clearly resolved by scanning tunneling microscopy (Fig. 3.17) [3.56]. The initial stages of Ag overlayer growth on the Au(111) have been studied by two groups recently

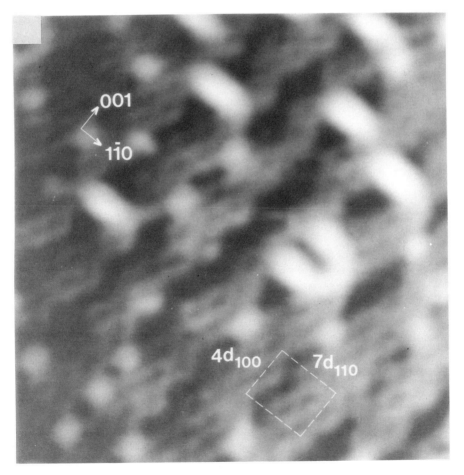

Fig. 3.17 120×120-Å^2 gray scale topograph of Au on the Ni(110) surface. A unit cell of the (7×4) structure is shown

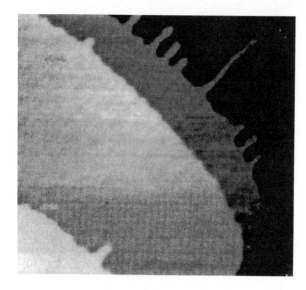

Fig. 3.18. $1100 \times 1100\text{-Å}^2$ gray scale image of Ag on the Au(111) with $V_{tip} = -200$ mV, and $I_{tunnel} = 0.1$ nA. From [3.65]

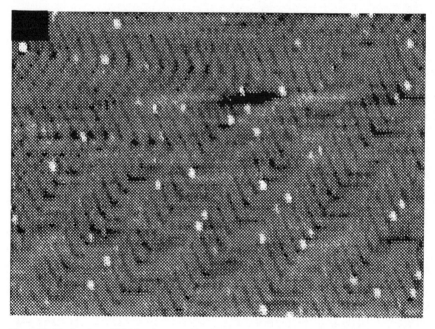

Fig. 3.19. Nucleation of Ni islands at specific sites on the Au(111) surface. The width of the image is 1900 Å. From [3.66]

[3.47, 57]. Since the lattice mismatch between Au and Ag is 0.24%, this growth is nearly homoepitaxial. Although the Au(111) substrate surface exhibits a $22-23 \times \sqrt{3}$ reconstruction, the Ag overlayer does not show this reconstruction. The Ag layer nucleates around defects, such as voids or step edges. At room temperature, the growth pattern shows a diffusion limited aggregation with local smoothing [Fig. 3.18]. From this study, the surface diffusion of the Ag overlayer was estimated. The early stage of Ni growth on the Au(111) surface shows unique nucleation sites [Fig. 3.19] determined by the Au substrate reconstruction [3.58]. At low coverage, the Ni overlayer forms a highly ordered metal island array on an atomically smooth surface.

Phase transitions and epitaxial growth can take place at temperatures from near 0 K to 2000 K. At present, the Curie temperature of the STM piezo elements limits the operation of an STM to < 300 °C. As shown in some examples, dynamics of surfaces can be successfully studied if it is slower than the STM scanning speed at room temperature.

3.4 Conclusion

Metal represents a particular challenge to scanning tunneling microscopy because of the small surface charge corrugation. Despite this, for the last decade scanning tunneling microscopy has contributed greatly to the understanding of metal surfaces, including structure, phase transitions, surface diffusion, adsorbate reaction and epitaxial growth. A UHV STM system in conjunction with other surface science techniques has become a powerful tool for the study of surfaces.

Acknowledgments. The author thanks P.J. Silverman, W.L. Brown, H.Q. Nguyen, F.M. Chua, Y. Hasegawa, R.S. Becker, R.J. Behm, D.D. Chambliss, S. Chiang, J.W.M. Frenken, P.K. Hansma, G.K. Gimzewski, R.J. Jaklevic, C.A. Lang, C.F. Quate, M. Salmeron, B.C. Schardt and S. Rousset.

References

3.1   G.A. Somorjai: *Chemistry in Two Dimensions Surfaces* (Cornell University Press, Ithaca 1981)
3.2   G. Binnig, H. Rohrer, Ch. Gerber, E. Weibel: Appl. Phys. Lett. **40**, 178 (1982); Phys. Rev. Lett. **49**, 57 (1982); Physica **109/110b**, 2075 (1982)
3.3   G. Binnig, H. Rohrer, Ch. Gerber, E. Weibel: Phys. Rev. Lett. **50**, 120 (1983)
3.4   G. Binnig, H. Rohrer, Ch. Gerber, E. Weibel: Surf. Sci. **131**, L379 (1983)
3.5   R.H. Fowler, L. Nordheim: Proc. Roy. Soc. London **A119**, 173 (1928)
3.6   R.D. Young, J. Ward, F. Scire: Phys. Rev. Lett. **27**, 922 (1971)
3.7   E.C. Teague: Ph.D thesis, North Texas State University (1978)
3.8   T. Engel: "Determination of Surface Structure Using Atomic Diffraction", in *Chemistry and Physics of Solid Surfaces V*, ed. by R. Vanselow, R. Howe, Springer Ser. Chem. Phys. Vol. 35 (Springer, Berlin, Heidelberg 1983)

3.9   Y. Kuk, P.J. Silverman, H.Q. Nguyen: J. Vac. Sci. Technol. A6, 524 (1988)
3.10 Y. Kuk, P.J. Silverman: Appl. Phys. Lett. 48, 1597 (1986)
3.11 L.W. Swanson, L.C. Crouser: J. Appl. Phys. 40, 4741 (1969)
3.12 H.W. Fink: IBM J. Res. Dev. 30, 461 (1986)
3.13 S.I. Park, J. Nogami, C.F. Quate: Phys. Rev. B36, 2863 (1987)
3.14 T. Sakurai, T. Hashizume, I. Kamiya, Y. Hasegawa, A. Sakai, J. Matsui, E. Kono,
      T. Takahaki, M. Ogawa: J. Vac. Sci. Technol. A6, 803 (1988)
3.15 K. Sugihara, A. Akira, Y. Akama, N. Shoda, Y. Kato: Rev. Sci. Instrum. 61, 81 (1990)
3.16 J. Tersoff, D.R. Hamann: Phys. Rev. Lett. 50, 25 (1983); Phys. Rev. B31, 805 (1985)
3.17 Y. Kuk, P.J. Silverman: Rev. Sci. Instrum. 60, 165 (1989)
3.18 Y. Kuk, P.J. Silverman: J. Vac. Sci. Technol. A8, 289 (1990)
3.19 J.K. Gimzewski, R. Moller: Phys. Rev. B36, 1284 (1987)
3.20 U. Dürig, J.K. Gimzewski, D.W. Pohl: Phys. Rev. Lett. 57, 2403 (1986); Bull. Amer. Phys. Soc.
      35, 484 (1990)
3.21 N.D. Lang: Phys. Rev. B36, 8173 (1987); Ibid 37, 10395 (1988)
3.22 R. Landauer: Z. Phys. B68, 217 (1987)
3.23 J.H. Coombs, J.K. Gimzewski: J. Microscopy 152, 841 (1988)
3.24 J.A. Kubby, W.J. Greene: J. Vac. Sci. Technol. B9, 739 (1991)
3.25 R.J. Hamers, R.M. Tromp, J.E. Demuth: Phys. Rev. Lett. 56, 1972 (1986)
3.26 R.S. Becker, J.A. Golovchenko, D.R. Hamann, B.S. Swartzentruber: Phys. Rev. Lett. 55, 2032
      (1985)
3.27 E.L. Wolf: Principles of Electron Tunneling Spectroscopy (Clarendon, Oxford 1985)
3.28 C.J. Chen: J. Vac. Sci. Technol. A6, 319 (1988)
3.29 N.D. Lang: Phys. Rev. B34, 1164 (1986)
3.30 P. Heimann, J. Heimanson, H. Miosga, H. Neddermeyer: Phys. Rev. Lett. 43, 1957 (1979)
3.31 M.P. Everson, R.C. Jaklevic, W. Shen: J. Vac. Sci. Technol. B9, 891 (1991)
3.32 Y. Kuk, L.C. Feldman, I.K. Robinson: Surf. Sci. 138, L168 (1984)
3.33 I.K. Robinson: Phys. Rev. Lett. 51, 1145 (1983)
3.34 J.R. Noonan, H.L. Davis: J. Vac. Sci. Technol. 16, 587 (1979)
3.35 D. Wolf, H. Jagydzinsler, W. Moritz: Surf. Sci. 88, L29 (1979)
3.36 Y. Kuk, P.J. Silverman, F.M. Chua: J. Microscopy 152, 449 (1988)
3.37 H.Q. Nguyen, Y. Kuk, P.J. Silverman: J. de Phys. 49, 7988 (1989)
3.38 J.C. Campuzano, M.S. Foster, G. Jennings, R.F. Willis, W. Unertle: Phys. Rev. Lett. 54, 2684
      (1985)
3.39 M.S. Daw, S.M. Foiles: Phys. Rev. Lett. 59, 2756 (1987)
3.40 V.M. Hallmark, S. Chiang, J.F. Rabolt, J.D. Swallen, R.J. Wilson: Phys. Rev. Lett. 59, 2879
      (1987)
3.41 Ch. Wöll, S. Chiang, R.J. Wilson, P.H. Lippel: Phys. Rev. B9, 7988 (1989)
3.42 J. Wintterlin, J. Wiechers, H. Brune, T. Gritsch, H. Hofer, R.J. Behm: Phys. Rev. Lett. 62, 59
      (1989)
3.43 Ph. Lippel, R.J. Wilson, M.D. Miller, Ch. Wöll, S. Chiang: Phys. Rev. Lett. 62, 171 (1989)
3.44 F.M. Chua, Y. Kuk, P.J. Silverman: Phys. Rev. Lett. 63, 386 (1989)
3.45 F. Jensen, F. Besenbache, E. Laegsgaard, I. Stensgaard: Phys. Rev. B41, 10233 (1990)
3.46 C.J. Chen: Phys. Rev. Lett. 65, 448 (1990)
3.47 D.D. Chambliss, R.J. Wilson: J. Vac. Sci. Technol. B9, 933 (1991)
3.48 G. Binnig, H. Rohrer, Ch. Gerber, E. Stoll: Surf. Sci. 144, 321 (1984)
3.49 R.J. Behm, W. Hösler, E. Ritter, G. Binnig: Phys. Rev. Lett. 56, 228 (1986)
3.50 J.K. Gimzewski, A. Humbert, J.G. Bednorz, B. Rehl: Phys. Rev. Lett. 55, 951 (1985)
3.51 N. Garcia, A.M. Baro, R. Garcia, J.P. Pena, H. Rohrer: Appl. Phys. Lett. 47, 367 (1985)
3.52 S.K. Sinha: Ordering in Two Dimensions, (North-Holland, New York 1980)
3.53 R.C. Jaklevic, L. Elie: Phys. Rev. Lett. 60, 120 (1988)
3.54 R.J. Schneir, R. Sonnenfeld, O. Marti, P.K. Hansma, J.E. Demuth, R.J. Hamers: J. Appl. Phys.
      63, 717 (1988)
3.55 J.W.M. Frenken, R.J. Hamers, J.E. Hamers: J. Vac. Sci. Technol. A8, 293 (1990)

3.56 Y. Kuk, P.J. Silverman, T.M. Buck: Phys. Rev. **B36**, 3104 (1987)
3.57 M.M. Dorek, C.L. Lang, J. Nogami, C.F. Quate: Phys. Rev. **B40**, 11973 (1989)
3.58 D.D. Chambliss, R.J. Wilson: To be published
3.59 R.S. Becker, J.A. Golovchenko, B.S. Swartzentruber: Phys. Rev. Lett. **55**, 987 (1985)
3.60 J.W. Winterlin, J. Wiechers, H. Brune, T. Gritsch, H. Höfer, R.J. Behm: Phys. Rev. Lett. **62**, 59 (1989)
3.61 D.D. Chambliss, R.J. Wilson: J. Vac. Sci. Technol. **B9**, 928 (1991)
3.62 J.K. Gimzewski, A. Humbert, J.G. Bednorz, B. Reihl: Phys. Rev. Lett. **55**, 951 (1985)
3.63 R.C. Jaklevic, L. Elie: Phys. Rev. Lett. **60**, 120 (1988)
3.64 J.W.M. Frenken, R.J. Hamers, J.E. Demuth: J. Vac. Sci. Technol. **A8**, 293 (1990)
3.65 M.M. Dovek, C.A. Lang, J. Nogami, C.F. Quate: Phys. Rev. **B40**, 11973 (1989)
3.66 D.D. Chambliss, R.J. Wilson, S. Chiang: Phys. Rev. Lett. **66**, 1721 (1991)

4. Adsorbate Covered Metal Surfaces and Reactions on Metal Surfaces

J. Wintterlin and *R.J. Behm*

With 22 Figures

Adsorption on metal surfaces is one of the major topics of surface science. Stimulated by the interest in a variety of technologically important processes such as catalytic reactions and corrosion, the interaction between adsorbates and metal surfaces has been studied for a long time. Phenomena such as adsorbate bonding, dissociation, surface diffusion, ordering processes, reactions with the surface or with other adsorbates, growth of three-dimensional layers of adsorbates or of reaction products, or finally desorption were studied by a variety of techniques. Except for a few methods such as low-energy electron microscopy (LEEM) [4.1] and field ion microscopy (FIM) [4.2] all of these techniques integrate over macroscopic surface areas. STM, on the other hand, provides local information on an atomic scale, both on the structure and on the electronic properties of the surface. This is particularly important for the investigation of e.g. the role of defects and impurities in surface reactions.

In the following, we present an overview of characteristic results of STM studies on these subjects. The following section deals with imaging of adsorbates on metal surfaces by STM. Based on experimental results obtained for individual adsorbates and on theoretical predictions, we discuss fundamental aspects of the representation of adsorbates in STM images. Imaging and resolution of adsorbates in close-packed adsorbate layers and first results on STM spectroscopy of adsorbates are also included in this section. In Sect. 4.2 we discuss STM studies on processes on metal surfaces such as adsorption, dissociation of adsorbed molecules, surface diffusion and ordering of adparticles. Investigations of mechanistic details and the role of surface defects in these processes represent central topics of STM investigations. This is equally important for Sect. 4.3, which reviews STM studies on those surface reactions where the structure of the metallic substrate is modified by the presence of the adsorbate. These include adsorbate-induced reconstructions of surfaces, which were the subject of a large number of STM studies. Oxidation reactions, where, in addition, the chemical state of the metal atoms is altered significantly, are also dealt with in this section. Finally, in Sect. 4.4 we discuss STM investigations on the epitaxial growth of metallic overlayers on metal substrates.

4.1 Imaging of Adsorbates by STM

4.1.1 Representation of Individual Adsorbates

Imaging of adsorbates by STM is most systematically studied in systems where isolated adsorbate particles are located inmid of flat substrate areas. In these cases the structure in the STM images, at the location of the adsorbed particles, directly reflects the impact of adsorbates. In most cases, however, the situation is less than ideal. The adsorbed particles either diffuse rapidly across the surface or, due to attractive interactions, cluster into islands where they are too closely spaced to be clearly resolved. At room temperature the mobility of most simple adsorbates is sufficiently high to prevent studies on individual, localized adsorbates. Therefore it would be desirable to perform such studies at reduced temperatures. Because of experimental difficulties, this is possible only with very few setups at present. For these reasons, only a few studies on individual adsorbates have been reported so far.

The potential of low temperature studies was demonstrated in a recent STM investigation of physisorbed Xe atoms, at 4 K, on a Ni(110) surface [4.3]. In the STM image in Fig. 4.1 five xenon atoms are resolved, which are represented as 1.6 Å high and 7 Å wide, cone-shaped protrusions on a flat surface. (The cluster of three atoms in Fig. 4.1 had been "assembled" by use of the tunnel tip.) Faint stripes represent the close-packed atomic rows of the Ni(110) substrate. It was found, by direct observation, that the xenon atoms bind to the nickel surface in

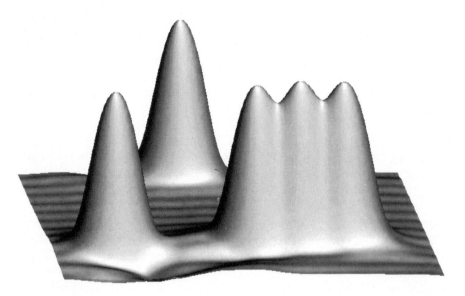

Fig. 4.1. Xe atoms physisorbed on Ni(110) at 4 K. Xenon atoms appear as protrusions, the rows of nickel atoms as light and dark stripes. From [4.3]

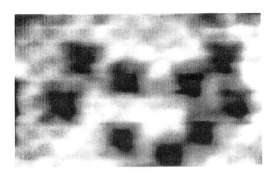

Fig. 4.2. Oxygen atoms (dark squares) adsorbed on Ni(100) (atomically resolved metal lattice, 28 Å × 17 Å). From [4.5]

the four-fold coordinated hollow sites. The maxima in the STM contours, which represent the Xe atoms, are somewhat lower than expected from a hard sphere model, which yields 2.7 Å. Nevertheless, Fig. 4.1 qualitatively corresponds to a naive picture of an STM contour as a representation of ion cores, directly reflecting a contour of charge density over the adsorbed Xe atoms [4.4].

This is no longer true in the example presented in Fig. 4.2 showing oxygen atoms adsorbed on a Ni(100) surface [4.5]. At 300 K, oxygen was found to be sufficiently localized on the time scale of the experiment to be imaged. The oxygen atoms are represented as quadrangular dark spots corresponding to holes of 0.3 Å in depth and 4 Å in width. The atomic lattice of the Ni(100) substrate is resolved as a weak periodic corrugation on the substrate areas. The oxygen atoms are located in the four-fold hollow sites. The four Ni atoms in their direct neighborhood lead to deviations from a round appearance of the O_{ad}-induced features and are responsible for the characteristic diamond-like shape and the 45° rotation with respect to the substrate lattice. Similar shapes would be expected from simple arguments on the electron charge distribution in front of the surface if we assume this to have its maxima above the Ni atoms. The appearance of pronounced minima at the locations of the O_{ad} apparently contradicts the well-known adsorption geometry of the O adatom. From structure analysis by LEED [4.6] and SEXAFS [4.7] it is known that the center of the oxygen atom is 0.8 Å above the topmost Ni layer. Hence, the apparent height of the O_{ad} in the STM image does not even qualitatively correspond to the geometric arrangement. The situation very much resembles that of semiconductor surfaces, where an understanding of the STM images is possible only by considering their electronic structure, while for clean metal surfaces STM images largely reproduce the topography of the surface [4.8, 9].

The correlation between electronic structure and tunnel current is illustrated in Fig. 4.3, which shows a schematic energy diagram of the tunnel junction between an adsorbate-covered metal surface and a metallic tip [4.10]. The density of states in front of the sample surface is indicated. For simplicity the tip density distribution is regarded as smooth and is omitted from Fig. 4.3. Only electrons from electronic states between the two Fermi levels in an energy window ΔE, which is defined by the tunnel voltage V_t, contribute to the net

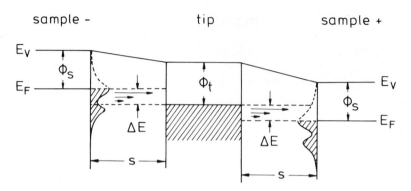

Fig. 4.3. Schematic energy diagram for the tunnel contact between a metallic tip and an adsorbate covered metal surface. Left junction: negative sample bias; right junction: positive sample bias. From [4.10]

tunnel current (I_t). In the left junction (sample negative) electrons tunnel from occupied states of the sample into empty states of the tip, and in the right junction (sample positive) they tunnel from occupied tip states into empty states of the sample. Due to the lower barrier height Φ seen by electronic states at the upper range of the energy window, these states carry the largest portion of the tunnel current (indicated by arrows). In addition, the characteristics of the respective wave functions play a role. The more delocalized s states contribute more to the tunnel current than the contracted d states of the same energy. Similarly, p_z orbitals will affect the tunnel current much more than p_x and p_y orbitals, which are concentrated in a plane parallel to the surface. Therefore the STM traces over an isolated adsorbed particle reflect the adsorbate-induced change in filled or empty state density in an energy window ΔE at E_F and are a function of the shape of the contributing wave functions.

According to the simplified interpretation of STM images by *Tersoff* and *Hamann* [4.8, 9], which is based on the "transfer Hamiltonian" formalism introduced by *Bardeen* [4.11], the tunnel current is proportional to the electron density of the sample, at E_F, at the position of the tip. Hence, the STM images represent contours of constant LDOS (local density of states) at E_F. For metals these resemble contours of the total electron density so that STM images of clean metal surfaces are closely related to their topography. Unlike the DOS of metal surfaces, the adsorbate-induced DOS may exhibit a pronounced spatial and energetic structure. Adsorbates can modify the LDOS near E_F by electronic states derived from adsorbate orbitals. Even if these are centered well outside the energy window ΔE, which is often the case, the state density at E_F may still be affected because of the broadness of the adsorbate resonances. In addition, the metal electronic structure can be changed by polarisation of metal electrons, removal of surface states, etc. Modifications of the barrier height Φ must also be considered. Variations in the decay of the wave functions in the barrier region may lead to images which depend on the vertical tip distance, i.e. on the tunnel current. Hence, the effects at the Fermi energy can appreciably deviate from the changes in the total DOS so that, in general, STM contours of adsorbates do not simply reflect the positions of their ion cores.

More quantitative predictions on the representation of adsorbates in STM images came from first calculations on adsorbate imaging by STM. *Lang* treated both electrodes in the jellium model and modeled the tip by placing a Na atom on one jellium surface. Different atoms adsorbed on the other surface were investigated. The tunnel current was calculated in a formalism equivalent to Bardeen's "transfer Hamiltonian" description. In the first publications the low voltage case was considered, restricting the contributing wave functions to states near E_F [4.12–14]. Figure 4.4a shows calculated traces of constant tunnel current over a sodium, sulfur and a helium atom adsorbed on the sample surface. It was found that these chemically very different atoms produced also different STM contours. S appears smaller than Na, which was attributed largely to the fact that the increase in state density at E_F for S is smaller than that for Na. In addition, the sulfur atom sits closer to the surface. Figure 4.4b shows the calculated state density for the Na/jellium and the S/jellium system. The Na $3s$ resonance, which is mostly above the Fermi level, still generates an appreciable increase of state density at E_F. For sulfur, where the $3p$ resonance is far below E_F, the contribution to the Fermi level state density is smaller. Hence, the increase of the tunnel current is smaller than would have been expected from the distance of the atom above the surface. The striking result that He, although far

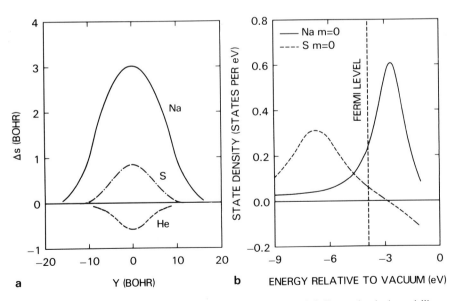

Fig. 4.4. (a) Calculated STM contours for sodium, sulfur, and helium adsorbed on jellium. **(b)** Calculated curves for the difference in eigenstate density between the metal adatom system and the bare metal. From [4.13]

above the surface, is imaged as a minimum was explained by the fact that its filled valence shell is far below E_F and does not contribute to the tunnel current. Its only effect is to polarize the metal electrons, producing a decrease in Fermi state density (not visible on the scale of Fig. 4.4b) and thus a reduction in tunnel current.

Another approach to a theoretical description of STM imaging of adsorbates was used by *Doyen* and coworkers [4.15, 16]. In their treatment the tunnel current was calculated "exactly", without making use of the "transfer Hamilton" operator, i.e. they allowed explicitly for interactions between the two electrodes. The wave functions of the tip/sample system were calculated by means of a semiempirical method. They modeled the tip by placing an adsorbed tungsten atom on a W(110) surface. Oxygen on Ni(100) [4.15] and on Al(111) [4.16] were studied in the framework of this model. In these calculations effects of the tip–sample distance were also found, which will be discussed further below.

We will now consider in more detail experiments for which the observation of individual adsorbates was reported. The resolution of Xe atoms on Ni(110) [4.3], at low temperatures, was already mentioned at the beginning of this chapter. Based on the above considerations, their appearance as protrusions must be correlated to variations in the electronic structure at E_F. In photoemission spectroscopy of xenon on Ni(110), the Xe $5p1/2$ and $5p3/2$ valence states were found to be located far below E_F, at -7.65 and -6.0 eV [4.17]. The lowest unoccupied states, the Xe $6s$ and $6p$ orbitals, were identified between 3.0 eV and 5.7 eV above E_F on Au(110) and Ru(0001) [4.18, 19] by IPES (inverse photoemission spectroscopy). It is not clear how the DOS at E_F is affected by these energetically remote states. In addition, calculations resolve a polarization of metal electrons, induced by the van der Waals bond to the Xe atoms which leads to a reduction of the metallic DOS at the location of the Xe_{ad} [4.20]. All these effects are relatively weak, rendering a straightforward interpretation of the experimental findings in terms of the electronic structure of Xe/Ni(110) difficult.

Even more complicated is the situation for adsorbed oxygen atoms. Individual oxygen atoms or small oxygen islands were observed on Ni(100) [4.5, 21], Al(111) [4.22] and Ru(0001) [4.23]. It was found that for not too large tunnel currents the oxygen atoms are imaged as holes by STM. Apparent depressions as deep as 1 Å (for Al(111) at $I_t < 1$ nA) were resolved. From results of previous LEED and SEXAFS studies, it is known that in all of these cases the oxygen atoms are located rather close to the surface, on the highly coordinated sites. The vertical distances amount to 0.6 Å for Al(111) [4.24], 0.8 Å for Ni [4.6, 7], and 1.2 Å for Ru [4.25]. (We do not consider a subsurface site, which has been postulated by some authors for Al(111)). Due to the high electronegativity of oxygen, there is a considerable charge transfer to the oxygen atom. The negative charge, however, is concentrated in the oxygen p orbitals, which appear at about -6 eV (p_z) and below (p_x, p_y) in UPS [4.26–28]. Hence, the valence shell of the O adatom does not contribute directly to the tunnel current. From his calculations for O adsorbed on a jellium surface, *Lang* postulated that the

O adatom affects the tunnel current predominantly by polarization of the metal electrons [4.29, 30]. In these calculations he found an accumulation of negative charge on the oxygen atom very close to the surface, surrounded by a hemispherical depletion zone inside the metal. This led to a distinct decrease in DOS at E_F, in addition to the p resonance at 7 eV below E_F, from which he predicted a small negative tip displacement of the order of 0.1 Å at the location of the O adatom, in qualitative agreement with the STM result. On the other hand, an O-induced *increase* in DOS above E_F was detected by IPES for O/Ni(100) [4.31]. Finally, *Doyen* et al. [4.15, 16] found in their calculations that the extended O 3s orbital, which is empty in the free atom, is partially occupied in the adsorbed O. This should lead to an increase in LDOS at E_F at large distances in front of the sample and consequently to protrusions in the STM trace at the location of the O_{ad}, for gap widths > 6 Å. On Al(111) O_{ad} produced likewise protrusions of ca. 0.3 Å height in these calculations. It is interesting to note that these authors found a negative tip displacement on Ni(100) for small gap widths, which at distances of > 6 Å turned into the positive displacement described above, i.e. in between there is a critical gap width where the STM is blind to the presence of the O_{ad}. At these close distances the contour lines of constant DOS at E_F also differed significantly from the calculated STM traces. The authors attributed these effects to contributions from the oscillatory inner part of the surface wavefunctions which are sampled by the W tip. At larger distances, where mainly the exponential tails of these wave functions are probed, there is at least qualitative agreement between these lines [4.15, 16]. As a whole, the interpretation of STM images of O_{ad} on metals still remains an open question.

Although in the case of O/Ni(100) later experimental data could not confirm the predicted contours, strong distance effects were observed in a number of adsorbate systems. For oxygen on Al(111), the O-induced holes in the STM image became less deep at close distances and additional central protrusions were observed at the location of each adatom [4.32]. Adsorbed C atoms, which were typically imaged as small protrusions, were found to become transparent to the STM at a certain distance [4.33] (Fig. 4.5). Finally, individual rows of the $(2 \times 1)O$ added row structure on Cu(110), which consist of linear chains of Cu and O atoms deposited on the Cu(110) substrate, appeared as indentations or protrusions depending on the gap width [4.34].

The formation of a chemical bond between an adsorbate and a metal substrate leads to changes in the electronic structure of the adsorbate and the neighboring surface metal atoms, which together form the adsorption complex. The local modification of the substrate atoms can also be detected in high resolution STM images. These images give a direct impression of the range over which surface chemical bonds act and over which distances neighboring adsorbates may influence each other via electronic interactions mediated through the substrate. Figure 4.5b [4.33] shows an STM image of three carbon atoms which appear as bright spots on an Al(111) surface. (We do not comment here on the enhanced atomic corrugation of clean aluminum [4.35, 36], which is not the subject of the present review.) The carbon atoms are located on the three-fold

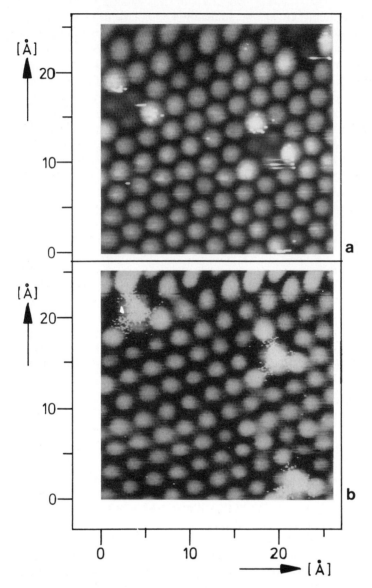

Fig. 4.5. STM images taken subsequently of almost the same Al(111) surface area, with two (**a**) and three (**b**) carbon adatoms respectively in the imaged area. (**a**) $V_t = -20$ mV, $I_t = 41$ nA; (**b**) $V_t = -70$ mV, $I_t = 41$ nA. From [4.33]

hollow sites, which, in addition, could be identified as hcp-type sites [4.33]. In Fig. 4.5a, recorded on the same surface area at slightly different tunnel parameters, the C adatoms appear to be transparent and the three nearest neighbor Al atoms, which are directly bound to the carbon, were resolved. These atoms appear darker, while the three next-nearest neighbor atoms appear brighter than the other Al atoms. This was attributed to adsorbate-induced changes in electron density rather than to vertical displacements of the respective aluminum atoms. A Friedel-type charge density oscillation induced by the electronegative C atom was proposed as an explanation for this effect. Comparable adsorbate-induced charge density oscillations were obtained in Lang's calculations, in which the metal was described in the jellium model. The STM images of C_{ad} on Al(111) reflect the size of the adsorption complex, which obviously reaches to next-nearest neighbors. A qualitatively similar result was obtained for a Si_{10} cluster deposited on a Au(100) surface, although without resolution of the individual substrate atoms [4.37].

STM imaging of larger, organic molecules such as Cu phthalocyanine on Cu(100) [4.38] is discussed in Chap. 7.

4.1.2 Resolution and Corrugation in Closed Adlayers

At higher coverages adsorbates form closed layers, which cover the entire surface or fractions thereof and which very often display an ordered arrangement of the adsorbed particles. Sometimes whole sequences of ordered overlayers are observed with increasing coverage. In these cases the adsorbed particles give rise to a corrugation pattern in STM whose amplitude differs from the height measured for individual adsorbed particles. The corrugation amplitude then depends on the instrumental resolution of the STM and on the characteristics of the adsorbate layer. Hence it is affected by the gap width, the sharpness of the tip and, most importantly, by the wave length of the corrugation, i.e. the distance between nearest adsorbate particles. Following the description of the instrumental resolution given by *Tersoff* [4.39] the lateral resolution for metal surfaces can be specified by a quantity $\sqrt{(R + s)/2\kappa}$, with κ being the inverse length, s the vertical tip–sample distance and R the tip radius. For a monoatomic tip and typical values for s and Φ, a lateral resolution of 2.5 Å is achieved. For a less perfect tip Fourier components of the surface corrugation on that length scale may still be detected, however with considerably reduced amplitude.

Hence, close-packed adsorbate overlayers are at the resolution limit of STM and, for a given lateral instrumental resolution, adsorbate particles in a closed adlayer are better resolved the larger their spacing. Correspondingly, in the first ordered adsorbate structure clearly resolved by STM – a mixed benzene/CO layer on Rh(111) [4.40] – resolution of the individual benzene molecules was assisted by CO "spacer" molecules. These data are discussed in Chap. 7.

The resolution of individual adsorbate particles in high resolution STM images of closed adlayers gives valuable structural information. This is particularly useful for the understanding of complex adlayer structures with large unit cells. This was demonstrated e.g. in an STM study on the $(2\sqrt{3} \times 2\sqrt{3})R30°$ superstructure of sulfur on Re(0001) [4.41]. Figure 4.6 shows a top-view image (a), together with a cross section along the marked line (b). The image displays a

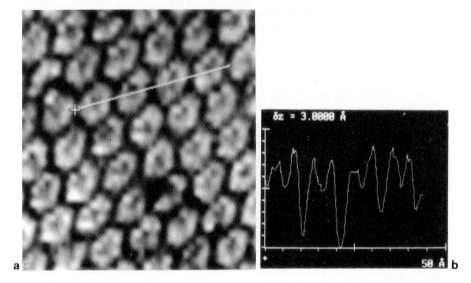

Fig. 4.6 (a): STM image $(65\,\text{Å} \times 62\,\text{Å})$ of the $(2\sqrt{3} \times 2\sqrt{3})\text{R}30°$-S overlayer on Re(0001). Bright rings: Hexagons of six sulfur atoms. **(b)**: cross section along the white line in part **(a)**. From [4.41]

hexagonal pattern of bright rings, each of which represents, according to the authors' interpretation, six S atoms arranged in a hexagon. The distances between the hexagonal rings of 9.5 Å correspond to the lattice constant of the $(2\sqrt{3} \times 2\sqrt{3})\text{R}30°$ overlayer. The troughs in between were found to be $1-2\,\text{Å}$ deep, while the central holes are more shallow and ca. 0.5 Å deep. The contrast between the sulfur atoms within the rings is even less. Qualitatively, these results are in good agreement with expectations from simple resolution arguments. The rings themselves are well resolved due to the larger separation between them. The smaller minima in the ring centers or between S atoms in the rings reflect the smaller distances between neighboring S atoms in these cases. Although the atomic structure of the substrate was not resolved by STM, the fcc-type three-fold coordinated site appeared most probable by analogy to S overlayers on other metal surfaces [4.42]. The STM data in addition revealed triangular distortions of the hexagons compatible with the overall threefold rotation symmetry of the overlayer. Based on these findings and on other results (Θ_S = 0.5 [4.43]) a model was derived for the $(2\sqrt{3} \times 2\sqrt{3})\text{R}30°$ structure, which consists of distorted S_6 rings with sulfur atoms in the fcc positions. Hence in this case STM results suggested a structural interpretation, which ruled out a previously proposed structure [4.43].

Resolution of the individual S atoms was reported also in a number of other STM studies on sulfur overlayers on metal surfaces, e.g. in a (2×1)S structure on

Mo(100) [4.44, 45], or in (2×2)S and $c(4 \times 2)$S structures on a stepped Cu(100) surface [4.46, 47]. (In part the data represent barrier height images which often gave better resolution.) Other examples of atomic resolution images of adsorbate layers include a sequence of iodine superstructure, a $(\sqrt{7} \times \sqrt{7})$I and two different (3×3)I overlayers on Pt(111)[4.48] and a number of ordered oxygen adlayers on different metal substrates. In the case of the iodine adlayers the experiments were performed under ambient pressure, using the barrier height imaging technique. Hence topographic corrugation amplitudes could not be specified. However, the images displayed details of a fine structure within the unit cells. Maxima in the Φ contours of different height were associated with iodine atoms sitting on different adsorption sites. A complete structure model, however, could not be established. Individual oxygen atoms in ordered layers were resolved for the (2×2)O structure on Ru(0001)[4.23] and for the (2×2)O and the $c(2 \times 2)$O on Ni(100)[4.5, 21]. Again, the more close-packed O atoms in the $c(2 \times 2)$ phase were much harder to resolve than the more distant adatoms in the $p(2 \times 2)$ phases. Finally, observation of a (2×2) hydrogen superstructure on Ni(111) was reported, which was correlated with a hexagonal pattern in STM images [4.49].

The above structures are characterized by interatomic distances which are larger than the substrate lattice spacings. Recently *Gritsch* et al. succeeded for the first time in resolving individual adsorbates in an adlayer which exhibits a density comparable to that of the substrate, in the (2×1)p2 mg structure formed by CO molecules on Pt(110) at coverages close to one monolayer [4.50]. Incidentally, this is also the first case where CO molecules could be resolved in a pure CO adlayer. Previous STM studies on CO adsorption had failed to resolve the individual CO molecules [4.51–53]. The presence of CO was detected indirectly by observation of a CO-induced structural transformation of the metal substrate (see Sect. 4.3.1). The inability to resolve the CO molecules was attributed to their high mobility at the low coverages present in those experiments. On the other hand, in the (2×1)p2mg structure the density of the CO molecules is so high that they are locked into an ordered layer. This resembles very much the case of the mixed $c(2\sqrt{3} \times 4)$ rect structure on Rh(111), where CO molecules are locked into a rigid lattice provided by the benzene molecules [4.54]. Figure 4.7a shows an STM image of the (2×1)p2mg-CO structure on Pt(110), which exhibits a characteristic zig-zag structure on the terraces. From results of other structural studies, it is known that the CO molecules are located on top of the metal atoms in the close-packed $[1\bar{1}0]$ rows of the Pt(110) surface [4.55]. Since the van der Waals diameter of CO of 3.1 Å is larger than the lattice constant of 2.77 Å of Pt along the $[1\bar{1}0]$ direction, the CO molecules are forced to tilt by about 20° in alternate directions from the surface normal (see model in Fig. 4.7b). The resulting zig-zag chains are clearly visible in Fig. 4.7a. Due to the very close packing of the molecules, the corrugation along $[1\bar{1}0]$ is very weak and became visible only – with an amplitude of 0.1 Å – for relatively high tunnel currents between 20 and 150 nA.

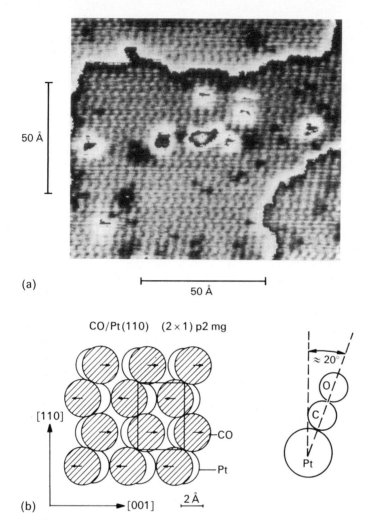

(a)

50 Å

50 Å

CO/Pt(110)   (2 × 1) p2 mg

[110]

[001]   2 Å

—CO

—Pt

≈ 20°

O

C

Pt

Fig. 4.7. (a) (2 × 1)p2mg-CO structure on Pt(110). Zig-zag chains represent rows of CO molecules, which are alternatingly tilted perpendicular to the rows of Pt atoms. From [4.50]. (b) Structure model of the (2 × 1)p2mg-CO structure. From [4.55]

4.1.3 Spectroscopy of Adsorbates

So far we have discussed topographic STM results on adsorbates assuming zero-voltage conditions, which limits the tunnel current to electronic states at E_F. Variations of the DOS with energy are probed by scanning tunneling spectroscopy (STS). In this mode of STM imaging, the size of the energy window $\Delta E = eV_t$ (see Fig. 4.3), which defines the energy range of contributing states is varied, and I_t is measured as a function of V_t. Changes in the tunnel current reflect the density of occupied and unoccupied states. In principle a complete, atomically resolved mapping of the DOS at the surface can be performed. However, spatially resolved spectroscopy has not yet been applied for the characterization of adsorbates as commonly as it is the case for semiconductor surfaces. This is expected to change in the future.

Integrated spectroscopy measurements on the DOS are usually performed by ultraviolet photoelectron spectroscopy (UPS) or inverse photoelectron spectroscopy (IPES). But in addition to the spatial resolution, STS differs from these techniques in several respects, which is particularly important for spectroscopy measurements on adsorbates. Firstly, spectra obtained by STM represent a convolution of sample and tip properties. This is not a major problem because the density of states of the (clean) metallic tip is expected to be rather smooth [4.56]. This effect can play a role, however, if during adsorption experiments the tip apex is also modified by an adsorbate. Special care has to be taken to maintain a spectroscopically well-defined tip for STS measurements. Secondly, for STS an energy-dependent weighting factor comes into play, which is not present in UPS or IPES. It results from the fact that the states at the upper edge of the energy window see a smaller potential barrier and therefore contribute more to the tunnel current than states at the lower edge. Hence, while V_t is increased during recording of the spectra, occupied and unoccupied states which enter the energy window ΔE will also have a different impact on changes in the tunnel current. Low-lying, occupied states will have little effect on the tunnel current and consequently on the STS spectra, while high-lying, unoccupied states will lead to stronger changes in the current. Therefore the sensitivity of STS rapidly decreases for measurements of occupied DOS with increasing energy below E_F. This effect is particularly important for spectroscopy on adsorbate-covered metal surfaces, since the characteristic adsorbate-related levels known from UPS are mostly well below E_F. Spectroscopy of unoccupied states, in contrast, is not hindered by this effect. Thirdly, STS measurements probe the DOS some ångstroms above the surface. UPS and IPES on the other hand integrate over several layers close to the surface. Therefore the sensitivity of STS will depend critically on the spatial extension of the respective wave function. Metal d states, because of their stronger contraction, are probed much less efficiently than the more delocalized s and p states. Finally, in systems displaying an electronic band structure, STS integrates over k space, while k-resolved measurements are possible by UPS or IPES. The integration over k space in STS, however, again involves a weighting factor which depends on the $k_{\parallel}$ component of the respective electronic state. The inverse decay length $\sqrt{(2m\Phi/\hbar^2) + (k_{\parallel} + G)^2}$ reduces the contribution of states with large $k_{\parallel}$ vectors to the tunnel current (G is a reciprocal lattice vector of the surface).

Early spectroscopy measurements with the STM were performed by recording dI_t/dV_t or $d\ln I_t/dV_t$ curves, at constant I_t, by use of a modulation technique [4.57]. Peaks in these spectra can be associated with the electronic states. Spectra on Ni(111)/(2 × 2)H obtained by this method displayed peaks in the occupied state density at -1.1 eV and at -1.8 to -2.2 eV, where the first one was correlated with a hydrogen surface state [4.49]. For a CO-covered

Ni(110) surface, the empty $2\pi^*$ state of adsorbed CO was identified by a peak around $+3.5\,\text{eV}$ [4.58], in good agreement with existing IPES data [4.59]. In contrast, the occupied states of CO_{ad}, the $5\sigma/1\pi$ states at $-9.2\,\text{eV}$ and the 4σ state at $-11.7\,\text{eV}$ below E_F [4.60], were not detected in these measurements. This method, however, suffers from simultaneous changes in the tip–sample distance as the tunnel voltage is swept and hence was replaced by fast measurements of I_t/V_t curves at fixed lateral and vertical positions of the tip. For this purpose the feedback loop is interrupted for short periods in which the spectra are recorded. As shown by *Feenstra* et al., plots of the normalized differential conductivity $(dI_t/dV_t)/(I_t/V_t)$ versus V_t are a good measure of empty state density, whereas occupied states are reduced by a factor which increases with energy below E_F [4.61]. The close resemblance of $(dI_t/dV_t)/(I_t/V_t)$ and the DOS was also confirmed in calculations [4.62].

The first experiments of this kind, with atomic resolution, were performed on metal/adsorbate systems. Figure 4.8 shows spectra from a Cu(110) surface, which were recorded in an energy range between -2.5 and $+2.5\,\text{eV}$ [4.63]. Part of the surface was covered with a (2×1) oxygen overlayer, which involves a reconstruction of the surface and will be discussed in greater detail in Sect. 4.3.1. This structure essentially consists of adsorbed chains of oxygen and copper atoms. Spectrum A was taken on the bare Cu(110) surface, spectra B and C above different points inside a (2×1)O island. On the clean surface two states, $0.8\,\text{eV}$ and $1.8\,\text{eV}$ above E_F, were resolved; no extra structures were found below E_F. The empty states were associated with two surface states which had been detected in angular-resolved IPES (KRIPES) measurements [4.64]. Interestingly, the two known surface states are located in the sp gap around the $\bar{Y}$ point in the two-dimensional reciprocal space so that, from their large $k_{\parallel}$ vector, only

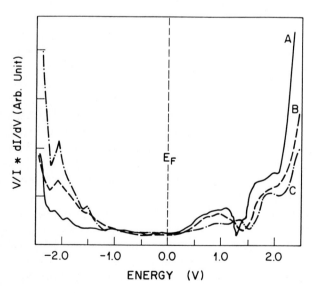

Fig. 4.8. Normalized tunnel conductances above clean Cu(110) (A) and above a (2×1)O island (B and C, from [4.63]

a small effect would have been expected in STM spectra, in contrast to the observations. The Cu 3d states between -2 and $-5\,eV$ known from angle-resolved UPS measurements [4.65] are weakened upon oxygen adsorption and could not be detected by STS, even on the clean surface, in agreement with our previous discussion. The same finding was reported for calculations of STS on an adsorbed Mo atom, which showed very little evidence of the Mo 4d state [4.66]. In the spectra recorded on the $(2 \times 1)O$ islands, the two peaks above E_F are shifted slightly to higher energy and additional density appeared at about $-2.0\,eV$. An occupied, antibonding state of oxygen, observed at about $-1.4\,eV$ in ARUPS and also detected by deexcitation spectroscopy of metastable atoms [4.67], was correlated with the increase in state density in spectra B and C at about $-2.0\,eV$. On the other hand, the main oxygen-derived spectral features – the O 2p-levels between -5.6 and $-7.8\,eV$ [4.65] – are far below E_F and are probably not detectable by STM. As a result, the differences between the clean and the oxygen-covered surface are relatively small in the energy range of Fig. 4.8.

These results illustrate the restrictions imposed on STM spectroscopy of adsorbates. In contrast to semiconductor surfaces, where the interest is concentrated predominantly on states around E_F, the electronic states of adsorbed atoms or small molecules are often well below E_F, which renders their spectroscopic identification by STM difficult. Clear adsorbate effects on the STM spectra were observed so far only in special cases, e.g. for Si_{10} clusters deposited on a Au(100) surface [4.68]. In this case, spectra recorded on top of a silicon cluster displayed a distinct band gap of about $1.0\,eV$ width. Hence, the Si clusters, although small and adsorbed on a metal surface, exhibit semiconductor behavior.

Topographic measurements, performed for a series of bias voltages, contain in principle the same physical information as STM spectra taken on different locations, and local I_t/V_t-curves can be constructed from a set of topographic scans recorded at different tunnel voltages. It has been shown that the apparent size of an adsorbed atom as a function of bias voltage reflects major features in the electronic spectra [4.66]. In his calculations, *Lang* found for $E > +1.0\,eV$ a reduced DOS in front of a sulfur atom adsorbed on a jellium surface concomittant with a predicted negative tip displacement. This correlation between apparent size of an adsorbate and its electronic structure failed only in the case of the 4d state of an adsorbed Mo atom. Because of the spatial contraction of the d electrons this state did not show up as a peak when the apparent height of a Mo atom was plotted against the tunnel voltage. Experiments on oxygen atoms on Al(111) showed, in accordance with the flat DOS around E_F [4.29], negligible changes of the apparent topography upon variations of V_t between -1.8 and $+1.0\,eV$ [4.32]. On the other hand, topographic images of the Cu(110)/$(2 \times 1)O$ structure acquired concurrently at -1 and $+2\,V$ displayed a distinct effect of the tunnel voltage [4.63]. Probing the occupied states, the full two-dimensional structure of this phase could be resolved, whereas for the empty states the atoms along one crystallographic direction were hardly resolved and the images merely showed sequences of parallel lines. This difference may reflect the spatial structure of the different wavefunctions.

In addition to the electronic structure, the vibrational structure of adsorbates is in principle also accessible by STM. Inelastic tunneling spectroscopy is commonly performed for molecules adsorbed on solid tunnel junctions [4.69]. Similar measurements by STM have been largely unsuccessful, despite extensive efforts [4.70]. Further details on this technique are given elsewhere [4.71].

4.2 Processes at the Metal–Gas Interface

4.2.1 Adsorption, Dissociation, Surface Diffusion

The resolution of individual adsorbates by STM opens new possibilities for the investigation of a large number of dynamic processes at the substrate – adsorbate interface and hence for the understanding of substrate–adsorbate interaction. In this section we will concentrate on those processes that leave the surface layer unchanged, such as adsorption, dissociation of molecules and surface diffusion. STM investigations can provide insight into local aspects of these phenomena, which have so far been studied almost exclusively by spatially integrating techniques. These local aspects involve, for example, the question of preferential sites for adsorption or dissociation, i.e. whether the reactivity of a surface for adsorption is determined by the properties of its ideal, single-crystalline parts or by defects (the concept of 'active sites' in catalysis [4.72]. A major drawback of STM in studying dynamic phenomena is its inability to detect very mobile particles. Furthermore, one has to make sure that the processes studied by STM are not affected by the measurement itself, e.g. by forces between tip and adsorbate through the electric field or by van der Waals interaction. The latter were made responsible for the controlled shifting of Xe atoms, by the STM tip, over a nickel surface [4.3].

Usually the adsorption process is too fast to be followed by STM. Under certain conditions however, STM measurements give details of this process, which was demonstrated recently for Xe on Pt(111) [4.73]. Following Xe adsorption at 4 K, most xenon atoms were found at steps or at point defects on the terraces, or they formed small clusters, although the thermal mobility of the adatoms at this temperature is negligible. The adatoms must have been able to travel over large distances after they first impinged on the surface. This was explained by assuming a mobile precursor, which has not yet fully accommodated to the surface. In contrast, similar experiments on Ni(110) identified mainly isolated Xe atoms, indicating a higher corrugation of the xenon–metal potential in the case of this surface. In addition, this example demonstrated the existence of heterogeneities in the metal–adsorbate potential. Surface defects or neighbored clusters of adatoms apparently stabilize the adsorption of Xe, and hence Xe adatoms are preferentially located at these sites.

Similar effects of surface heterogeneities in the substrate adsorbate potential are also frequently observed in adsorption experiments at $T \geqslant 300\,K$ where most other STM measurements were performed. For example hexatriene, which at room temperature adsorbed as an intact molecule on an Al(111) surface, clearly preferred defect sites such as step edges and preadsorbed C atoms on the terraces [4.74]. This indicates a higher adsorption energy at higher coordination or defect sites, together with sufficient mobility for migrating over clean, defect-free terraces at this temperature. On the other hand, for oxygen adsorption on Ru(0001) [4.23], Ni(100) [4.5], and Al(111) [4.22], the adatoms were found to be uniformly distributed over the terraces with no preferences for defect sites, pointing to minor variations in the site-specific adsorption energy and/or to a reduced mobility of the adatoms. In addition, the adsorbate–adsorbate interactions lead to changes in the (effective) metal adsorbate potential, which is reflected e.g. by the formation of oxygen islands observed for these systems. These interactions generally reduce the mobility of adparticles.

Steps and other surface defects can also be active for other surface processes. A distinct effect of atomic steps was detected by STM in the case of ethylene adsorption on the hexagonally reconstructed Pt(100) surface [4.51, 52]. It was observed that the hex $\rightarrow (1 \times 1)$ structural transformation of the substrate (see Sect. 4.3.1), if induced by $C_2H_{4,ad}$, starts at step sites in contrast to the same reaction initiated by CO_{ad} or NO_{ad}. This finding was explained by assuming some dissociation products C_2H_n rather than $C_2H_{4,ad}$ to be the active species in the transformation reaction. From earlier experiments it was known that, at elevated temperatures, adsorbed ethylene rearranges and loses part of its hydrogen. Furthermore, the decomposition rate of adsorbed hydrocarbons was found to be higher on stepped rather than on flat surfaces [4.75]. Based on these results the STM data could be interpreted in terms of a simple model. The adsorbed ethylene dissociates rapidly at atomic steps of the reconstructed Pt surface. The resulting, very reactive species then induces the hex $\rightarrow (1 \times 1)$ phase transformation close to the locations where they have been formed. Further growth of the (1×1) phase occurs at the perimeter of the carbon-covered transformed islands. Thus, although the adsorbate itself could not be resolved, the STM results revealed clear evidence for a defect-induced dissociation by observation of its direct aftermaths.

Surface diffusion of adsorbed particles, and in a number of cases also self-diffusion, are essential steps to most adsorption processes and surface reactions. Comparable to chemical reactions, surface diffusion often involves a complex sequence of different microscopic processes, all of which contribute to the macroscopic diffusion constant. The complex nature of these processes becomes evident from the wide range of values found for this quantity. STM measurements, in contrast to integral measurements, offer the chance to discriminate between the different contributions to surface diffusion and hence may become very valuable for its understanding.

For technical reasons, however, diffusion measurements are currently limited to a narrow range of mobilities. For direct observation, diffusion processes should not be faster than the time scale for image acquisition, mostly on the order of seconds to tens of seconds. As mentioned above, for instance, it was not possible to image isolated CO_{ad} or NO_{ad} molecules on Pt(100) because of their high mobility. On Ru(0001), individual oxygen atoms are mobile at 300 K, but diffuse sufficiently slowly that they can still be observed by STM [4.23]. The STM images indicate that the mobility of the O adatoms is reduced by adsorbate–adsorbate interactions. Vacancies in an ordered island are sufficiently immobile that their "hopping" occurs on a timescale which is comparable to the frequency of STM images. Therefore individual hops of adatoms can be followed from image to image. This is illustrated in the three images in Fig. 4.9, which were recorded subsequently on the same area of an oxygen-covered Ru(0001) surface ($\Theta_o \approx 0.22$). The four vacancies located near the center of the images remain on their sites in the first two patterns. In the third pattern, two of these vacancies have moved by one site. The vacancy at the bottom has changed its site from pattern to pattern. From an extended series of such measurements, a typical hopping frequency of 1–2 hops per vacancy and 100 s was estimated. By use of the Einstein relation this can be converted into a diffusion coefficient, yielding $D \approx 10^{-17}\,\mathrm{cm^2\,s^{-1}}$.

Another procedure to determine diffusion rates with the STM was described by *Binnig* et al. [4.76]. They analyzed random spikes in the tunnel current, which were associated with adatoms rapidly passing through the tunnel cone. A diffusion constant of $6 \times 10^{-12}\,\mathrm{cm^2\,s^{-1}}$ was obtained for oxygen atoms on Ni(100) at 350 K. This method is applicable for more mobile adsorbates, which are too fast for direct imaging.

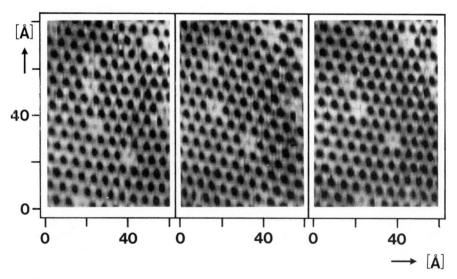

Fig. 4.9. Sequence of STM images taken of about the same area on an oxygen-covered Ru(0001) surface, demonstrating the mobility of vacancies in the adlayer ($\Theta = 0.22$, from [4.23])

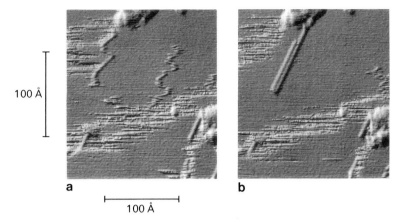

Fig. 4.10a,b. STM images recorded consecutively on the same area of a Cu(110) surface with a low amount of adsorbed oxygen. Two isolated, mobile (2×1)O strings in (**a**) are seen to be stabilized by condensation in (**b**). From [4.79]

Lateral interactions within adsorbate islands may be so strong that the probability for single particle motion becomes smaller than that for motion of entire islands. This was observed for islands of the reconstructed (2×1)O phase on Cu(110) [4.34, 77–79]. These islands are formed by long chains of oxygen and copper adatoms which, for small O coverages, exist as isolated strings on the bare surface. This is shown in Fig. 4.10a where the two thin, zig-zagged lines on the central terrace represent single –O–Cu–O– chains of almost 100 Å length. The irregular shape in the STM image results from dynamic effects rather than reflecting the real geometry of the chains. From scan to scan the positions of the chains changed leading to the observed features. These data indicate strong one-dimensional bonds within the –O–Cu–O– chains, which leave them intact while they diffuse across the surface. From these images it cannot be determined whether this migration takes place in a series of correlated jumps or whether it represents a true simultaneous process. The latter however, appears rather implausible. In Fig. 4.10b the two strings are seen adjacent to each other and appear stable, having lost their mobility. Obviously additional attractive interactions exist between adjacent chains, reducing their mobility if they come close to each other.

The effects of adatom–adatom interactions are most pronounced for diffusion of metallic adsorbates, including self-diffusion on metal surfaces. This was investigated in recent STM studies on Au(111) [4.80, 81] and Cu(110) [4.79]. Different from measurements by FIM, where hopping rates of isolated atoms on small terraces are obtained, these STM results yielded information about mass transfer. This includes the fast motion of atoms across flat terraces, the slower motion along step edges and finally, the much slower release of atoms from step edges. In the case of Cu(110) the mobility of atoms along the steps was found to be relatively high. This process leads to a rapid meandering of steps at room temperature, which gives them a "fringed" appearance in the STM images. Note the broad, streaky bands in Figs. 4.10a and b, representing step edges, which change their position from scan to scan. The motion of metal atoms across the terraces is very fast in all cases and could not be detected so far by STM due to the lack of cooling facilities in most experiments.

4.2.2 Formation of Ordered Adsorbate Layers

Ordering processes in adsorbate overlayers have long been studied by diffraction methods, such as LEED (SPALEED – spot profile analysis LEED) [4.82]. Again, these methods give quantitative, but integrated information on the long-range order and on the distribution of defects, but are not able to resolve local phenomena in these processes. Hence, STM measurements, which provide this information on an atomic scale, are ideally suited to complement data from diffraction studies. These local phenomena comprise the growth of adsorbate islands and the creation of point defects and dislocations in the adlayer. Although the STM measurements were mostly not performed in situ, during the adsorption process, "snapshots" recorded subsequently gave information on the kinetics of surface phase transformations and on the stability of metastable structures such as small islands and surface defects.

Antiphase domains in closed, ordered layers occur in almost all overlayer structures because of the reduced translational or rotational symmetry as compared to the substrate. In STM studies they were observed rather frequently: e.g. for sulfur layers on Re(0001) [4.41] and Mo(100) [4.44] and for mixed benzene/CO layers on Rh(111) [4.40, 54]. The presence of different domains on perfectly ordered substrate terraces is indicative of kinetic restrictions in the adlayer formation process, which prevent the removal of boundaries between large domains. Sometimes correlations between atomic steps on the substrate and the orientation of domains were observed. For an S overlayer on a stepped Cu(100) surface, steps of different directions were found to evoke different overlayer symmetries [4.46, 47]. Domains of (2×2)S and $c(4 \times 2)$S domains were observed simultaneously on the same terrace, if limited by steps of variable orientations. Point defects were also observed, e.g. vacancies in the (2×2)O structure on Ru(0001) [4.23] and distorted structural units in the $(2\sqrt{3} \times 2\sqrt{3})$R30°-S overlayer on Re(0001) [4.41].

A systematic STM study on ordering processes in an adlayer was reported for oxygen adsorption on Ni(100) [4.5]. Previous investigations on Ni(100)/O had agreed on the formation of (2×2) and $c(2 \times 2)$ phases with ideal coverages of 0.25 for the (2×2) and 0.50 for the $c(2 \times 2)$ [4.27]. The actual coverages and the adlayer order, however, had been a matter of debate. Figure 4.11a shows the Ni(100) surface after adsorption of oxygen at 410 K. Diamond-shaped black features represent oxygen atoms adsorbed on hollow sites, the grey background granulation stems from the atomically resolved, bare Ni(100) surface. The STM

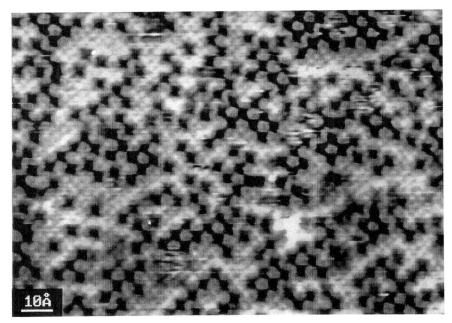

(a)

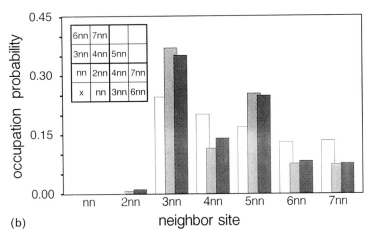

(b)

Fig. 4.11. (a) STM image of adsorbed oxygen ($\Theta_O = 0.16$) on Ni(100), showing local ordering into (2×2) islands. (b) Occupation probabilities for the nth neighbor sites for random adsorption with exclusion of nn and 2nn sites at $T = 0$ K (white bars), for adsorption at 410 K (image a, grey bars), and for adsorption at 370 K (black bars). From [4.5]

images were recorded at $T = 300$ K and sporadic adatom hops between subsequent images were observed. From the observed hopping rate a diffusion constant of $D = 5 \times 10^{-20}$ cm^2 s^{-1} was derived. (The number quoted in [4.76] for the same system was determined for a higher temperature.) The oxygen atoms appear to be almost randomly distributed over the surface. Upon closer inspection, however, small clusters with the oxygen adatoms arranged in a (2×2) lattice can be identified. A quantitative analysis revealed that the oxygen coverage is 0.16, i.e. below that of the ideal (2×2) phase, and that the mean (2×2) island size is only three atoms. The probability for occupation of the nth nearest neighbor sites of the adsorbed oxygen atoms was taken as a quantitative measure of the adlayer order. The numbers for this quantity obtained from Fig. 4.11a are shown as grey bars in Fig. 4.11b, together with a matrix giving the notation of neighbor sites around an atom (x). For a perfectly ordered (2×2) structure only 3nn, 5nn, 9nn etc. nearest neighbor sites would be occupied. A completely random distribution with exclusion of nn and 2nn sites would lead to the distribution represented by the white bars in Fig. 4.11b. The actual distribution reveals a significantly higher population of 3nn and 5nn sites than obtained for a random distribution, in accordance with the qualitative impression of small (2×2) islands in Fig. 4.11a. The considerable occupation of 4nn, 6nn etc. sites, which would not be occupied in a perfectly ordered (2×2) phase, reflects the many antiphase domain boundaries visible in Fig. 4.11a. From the fact that no atoms were observed to reside on nn sites and only a few on 2nn sites, strongly repulsive interactions between atoms on these sites were deduced. Correspondingly, the higher occupation probability for 3nn sites as compared to a random distribution points to weak attractions between adatoms on these sites. The black bars in Fig. 4.11b represent data obtained from a similar experiment, where adsorption was performed at a slightly lower temperature ($T = 370$ K). The distribution is very similar to that of the 410 K experiment, i.e. the adlayer order is not significantly affected by the temperature. From these findings the authors concluded that Fig. 4.11a represents the adlayer at $\Theta_o = 0.16$ in its thermodynamic stable configuration rather than a metastable phase stabilized by kinetic restrictions.

4.3 Structure Modifications of Metal Surfaces

4.3.1 Adsorbate-Induced Reconstructive Transformations

Surface reconstructions, although not as ubiquitous as for semiconductors, are an important phenomenon for a number of metals. The (110) surfaces of most fcc transition metals and the (100) planes of the 5d fcc transition metals Ir, Pt and Au are known for their instability towards reconstructions [4.83, 84]. In all of these cases, the stability of a certain surface geometry can be crucially affected by the presence of adsorbates. Hence, adsorption or desorption processes are often accompanied by a structural transformation of the surface, either a reconstruction or, if the clean surface is reconstructed, a removal of the reconstruction. The (1×2) and (1×3) missing-row reconstructed Au(110) [4.85] and the (2×1)O reconstructed Ni(110) surface [4.86] were the first metallic systems which were studied by STM. They were followed by a number of further STM studies on systems exhibiting adsorbate induced structural transform- ation. These include the interaction of different adsorbates with the hexagonally reconstructed Pt(100) surface [4.51–53, 87, 88], the formation of the streaky (1×2)H phase on Ni(110) [4.89–91], CO adsorption on the (1×2) missing-row reconstructed Pt(110) surface [4.50, 92, 93], the oxygen-induced formation of the Cu(110)/(2×1)O structure [4.34, 63, 77–79, 94], the $c(6 \times 2)$O [4.95, 96] reconstruction of Cu(110), the $(\sqrt{2} \times 2\sqrt{2})$O structure on Cu(100) [4.97] and finally the K-induced reconstructions into $(1 \times n)$K phases on Cu(110) [4.98].

In all of these systems the two-dimensional density of topmost layer substrate atoms in the reconstructed phase differs from that of the bulk-like surface. Hence, the structural transformation is connected with a change in substrate atom density and consequently with a major rearrangement of the surface topography. In the hex → (1×1) transition on Pt(100), which is initiated by adsorption of CO, NO or C_2H_4, the structural transformation is connected with a density reduction of 0.2 monolayers of Pt atoms in the topmost layer. From STM studies it was found that the excess atoms condense into small (1×1) "islands" on top of the former topmost layer [4.51–53, 88]. By monitoring their time-dependent formation, the initiation and progress of the reaction could be followed. A complex nucleation and growth scheme was developed from these data.

If the mobility of substrate atoms is low, the reaction and the resulting surface topography can be completely determined by surface migration effects. This was illustrated in a study on the CO-induced Pt(110) $(1 \times 2) \to (1 \times 1)$ transition [4.92, 93]. The clean Pt surface exhibits a missing-row reconstruction, which is removed by adsorption of CO. The transition corresponds to a change in density of Pt atoms from $\Theta_{Pt}(1 \times 2) = 0.5$ to $\Theta_{Pt}(1 \times 1) = 1.0$. Sequences of STM images recorded during CO adsorption at 300 K showed that under these conditions the transformation proceeds via formation of small, crater-like features. Their centers are formed by small (1×1) areas, covering 6 to 8 Pt atoms and surrounded by a rim of Pt atoms. With progressing reaction, these features grow in number but not in size, until at the end they cover the entire surface. Their formation results from kinetic restrictions. At 300 K Pt adatoms on this surface are almost immobile. The formation of these (1×1) structures requires a transport of 4–6 Pt atoms over a few lattice sites only.

The complex, reconstructive interaction of oxygen with Cu(110), which has been investigated by a variety of different techniques including STM [4.34, 63, 77–79, 94–96], shall be discussed in more detail in order to demonstrate the potential of STM studies for achieving a mechanistic understanding of surface processes on an atomic scale.

Oxygen is known to adsorb dissociatively at 300 K and to form two ordered structures:

$$Cu(110) \xrightarrow{10 \, L \, O_2} (2 \times 1)O \xrightarrow{10^4 - 10^5 L \, O_2} c(6 \times 2)O \ .$$

For both the (2×1)O and the (6×2) phases, surface reconstructions were assumed. However, despite numerous investigations, structure models for these phases remained under discussion and very little was known about the transformation mechanisms. For the (2×1) phase, two models were discussed, a missing-row (e.g. [4.99]) and a buckled-surface reconstruction (e.g. [4.100]). Both are formed by strings of Cu and O atoms in [001] direction, with $\Theta_O = 0.5$. In addition, the missing-row structure implies the removal of every other row of Cu atoms in [001] direction, whereas in the buckled-surface model these rows are still present, but are displaced vertically relative to the neighboring –O–Cu–O– strings.

Following oxygen adsorption at 300 K the formation of characteristic, strongly anisotropic islands was observed by STM [4.34, 63, 77–79, 94], resolved as dark features on the flat terrace in Fig. 4.12 [4.79]. The islands consist of parallel streaks in [001] direction; their spacings of 5.1 Å correspond to the two-fold substrate lattice constant in [1$\bar{1}$0] direction. Under certain conditions an additional corrugation, with a wavelength of 3.6 Å equal to the lattice constant in [001] direction, could be resolved along these streaks. Hence these protrusions are arranged in the (2×1) lattice characteristic for the lower coverage phase. The STM contours, however, were found to strongly depend on the experimental tunnel conditions. Often the [001] corrugation could either not be resolved, leading to a one-dimensional appearance of the streaks, or the contours were even reversed as shown in Fig. 4.12. To understand the physical origin of these differences, effects of the electronic structure [4.34, 63] and of adsorbates on the tip [4.78] were considered, but a satisfying explanation has not yet been given. Hence, the STM contours of the (2×1) phase could not

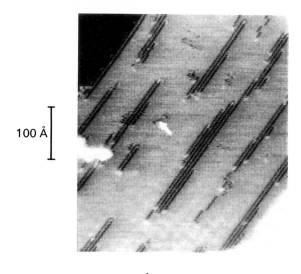

100 Å

100 Å

Fig. 4.12. STM image of (2×1)O islands (dark streaks) on Cu(110). From [4.79]

simply be related to atomic positions and, as a consequence, could not be used to discriminate between the two structure models described above.

More information on both structural and mechanistic aspects of the $(2 \times 1)O$ phase and its formation was derived from monitoring the process of (2×1) formation by STM [4.34, 77–79, 94]. From the uniform distribution of the (2×1) islands on the flat terraces, a homogeneous nucleation scheme was deduced. The minimum size for stable clusters at room temperature (critical nucleus) was estimated from these observations to be a single string of about 6 Cu and O atoms. The elongated shape of the islands reflects a strong anisotropy in the growth rates along [001] and [1$\bar{1}$0] directions, which was explained by differences in the interaction energies within and between these rows. Following these ideas, strong chemical bonds act within the –O–Cu–O– strings, along [001], while interactions between them, in [1$\bar{1}$0] direction, are much weaker. Monte Carlo simulations using realistic values for the interaction energies reproduced a growth mode similar to that observed in the STM images [4.78]. This result indicates that, in contrast to expectations, the difference in surface diffusion between the directions parallel and perpendicular to the close-packed rows on the substrate does not play a crucial role. Finally, STM images of smaller (2×1) islands often indicated surface mobility of entire –O–Cu–O– strings, which was described in detail in Sect. 4.2.1. In Fig. 4.12 it is reflected by the irregular shape of single isolated strings (one is marked by an arrow) and by the segmented appearance of a single row, which is part of an island on the right hand side. Lateral jumps of entire strings were also made responsible for the removal of antiphase domain boundaries between (2×1) islands [4.79].

As mentioned above, STM imaging of the dynamic progress of the structural transformation also gives information on structural aspects. The two (2×1) structure models exhibit different two-dimensional densities of Cu atoms in the topmost layer ($\Theta_{Cu, missing\ row} = 0.5$, $\Theta_{Cu, buckled\ surface} = 1.0$). Therefore it must be possible to discriminate between them by the observation of mass transport during (2×1) formation, or its absence. Figure 4.12 does not show any deposition of Cu atoms released from the nearby (2×1) islands, which could have been expected for a missing-row structure. For a systematic investigation, large scale in situ adsorption experiments were performed. Figure 4.13 shows a series of STM images of the same area recorded during oxygen adsorption [4.78]. The oxygen coverage increased from $\Theta_O = 0$ in Fig. 4.13a to $\Theta_O = 0.4$ in Fig. 4.13d. Comparison of the step terrace topography in the four images reveals that upon formation of the (2×1) islands (imaged as faint bright streaks) terraces are eroded. Hence the formation of the $(2 \times 1)O$ phase is connected with a consumption of Cu atoms, which are transported from monoatomic steps to the (2×1) islands on the terraces. In a later stage, (2×1) islands adsorbed on the terraces block further terrace erosion by stabilizing the steps. This explains why the steps lose their "fringed" appearance (which indicates fluctuations due to diffusion of Cu atoms along the steps, see Sect. 4.2.1) and become linear in the course of the phase transformation. The same stabilization effect is also responsible for the increasing alignment of steps along [001], along the direction of the (2×1)

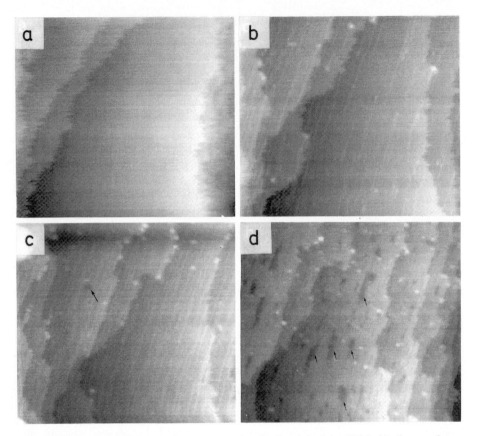

Fig. 4.13a–d. Oxygen adsorption on Cu(110), in situ monitored by STM. (a) clean surface, (b) $\Theta_O = 0.06$, (c) $\Theta_O = 0.17$, (d) $\Theta_O = 0.40$. Arrows indicate formation of monolayer deep holes on the terraces. From [4.78]

islands. When in a later stage of the transformation most of the steps are fixed by (2×1) islands along the steps, additional Cu atoms must be removed from the flat terraces. One-layer-deep holes are formed in the topmost Cu layer, which appear for the first time at $\Theta_O = 0.17$ in Fig. 4.13c (see arrows). The $(1 \times 1) \rightarrow (2 \times 1)$ transformation thus results in a completely modified surface topography. It also strongly affects the stability of step edges and hence the mobility on the surface.

These results suggest an *increase* of the copper coverage in the (2×1) islands, which is neither compatible with a missing-row nor with a buckled-surface reconstruction since these imply a decrease or no change, respectively, in the density of surface Cu atoms. However, an added-row reconstruction mechanism [4.34], as depicted in Fig. 4.14, could explain the experimental findings. This model is based on the high mobility of Cu surface atoms discussed in Sect. 4.2.1. Cu atoms can "evaporate" from step edges and move rapidly across the flat

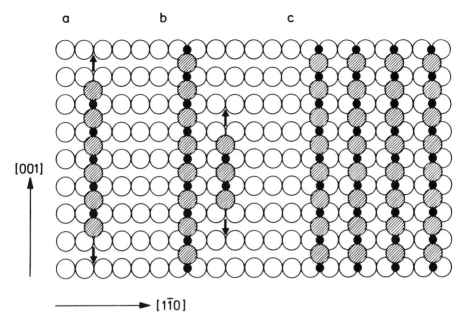

Fig. 4.14a–c. Atomistic model of the added-row model of the $(2 \times 1)O$ reconstruction of Cu(110). Filled circles: O atoms; shaded circles: Cu atoms on top of the substrate atoms (open circles). **(a)** Single strings of Cu–O adatoms; arrows indicate preferential growth direction. **(b)** Nucleation of a neighboring added row. **(c)** Two-dimensional island of the $(2 \times 1)O$ added-row phase. From [4.34]

terraces. At the same time oxygen atoms, resulting from O_2 adsorption and subsequent dissociation, are present on the terraces. Both species are not observed by STM because of their high mobility. From this mixed phase of mobile adparticles short strings of Cu and O atoms can condense on the terraces to form (2×1) nuclei (Fig. 4.14a). These grow rapidly along [001] and much more slowly in [1$\bar{1}$0] direction (Fig. 4.14b). This mechanism represents a two-dimensional analog to the precipitation of a solid from a "fluid" phase [4.34]. The activated step in this phase transformation, i.e. the release of Cu atoms from steps or, in later stages, from terrace sites, and the nucleation of (2×1) islands are spatially uncorrelated. Finally, upon completion of the structural transformation at $\Theta_O = 0.5$ (Fig. 4.14c), the added-row structure is identical to the missing-row reconstruction, in good agreement with results from earlier studies on the structure of this phase. However, the formation mechanism is different from a missing-row structure, and also the geometric level of (2×1) islands with respect to the bare metal is different.

Exposing the (2×1) reconstructed Cu(110) surface to much higher amounts of oxygen at $T \geq 300\,\mathrm{K}$ initiates a second structural transformation and a $c(6 \times 2)$ structure is formed. For this higher coverage phase a reconstruction was also assumed, but its actual structure was much more uncertain than that of the (2×1) phase. STM images of the $c(6 \times 2)$ structure showed a two-dimensional

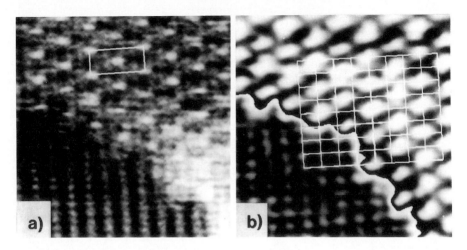

Fig. 4.15. (a) STM topograph showing coexisting domains of c(6 × 2) phase (*top*) and (2 × 1) phase (*bottom*); a c(6 × 2) unit cell is indicated. (b) Same area; the net indicates the relative registry of the c(6 × 2) with respect to the (2 × 1) structure. From [4.95]

corrugation pattern consisting of a quasi-hexagonal arrangement of round protrusions [4.95, 96]. Figure 4.15b [4.95] shows a c(6 × 2) domain adjacent to a (2 × 1) domain, the latter being identified by its row structure. Higher resolution images of the c(6 × 2) structure, such as the one shown in Fig. 4.15a, displayed additional weaker features between the large "bumps". The c(6 × 2) unit cell is marked in this figure. These high-resolution images pointed to a more complex structure, presumably with two different adsorption sites for the oxygen atoms as predicted from vibrational spectroscopy [4.101]. The resolution and representation of the fine-structure details varied in the experiments, comparable to observations on the (2 × 1) structure, whereas the larger protrusions were qualitatively not affected by the experimental parameters. Additional structural information came from STM images recorded in the coexistence region of the (2 × 1) and c(6 × 2) phases [4.95, 96]. These images showed that the round protrusions represent independent structural units, which can exist as isolated features in (2 × 1) reconstructed regions. These features displayed some mobility, predominantly along boundaries between c(6 × 2) and (2 × 1) domains. This indicates that they represent single atoms or very stable clusters of atoms. On the other hand, because of their low density, these protrusions cannot simply correspond to the O adatoms in this high coverage phase. In addition, from the lattice drawn in Fig. 4.15b, the position of the c(6 × 2) structure relative to the (2 × 1) reconstructed phase was obtained. However, even with this additional information the structure of the c(6 × 2) phase could not be unambiguously derived from STM images.

For more information the progress of the (2 × 1) → c(6 × 2) transformation was investigated by STM [4.95, 96]. Figure 4.16 [4.96] shows an image recorded in the early stages of the transformation, where most parts of the terraces are still

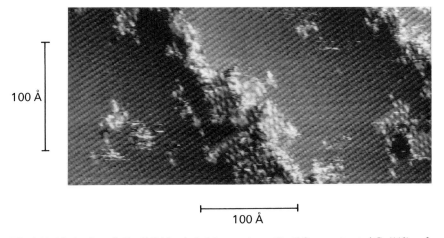

100 Å

100 Å

Fig. 4.16. Nucleation of c(6 × 2)O islands (bright spots) on a (2 × 1)O reconstructed Cu(110) surface. From [4.96]

covered by the typical –O–Cu–O– chains of the (2 × 1) structure. Additional bright spots are resolved, which are concentrated near step edges. Most of them cluster together, and small areas display a hexagonal arrangement of these features. Hence they are identical to the protrusions in the ordered c(6 × 2) phase. The large amount of disorder visible in Fig. 4.16 results from the lower temperature during adsorption (300 K), while the experiment in Fig. 4.15 was performed at 370 K. The concentration of c(6 × 2) islands along step edges (e.g. in the image center of Fig. 4.16) indicates prevailing heterogeneous nucleation, in contrast to the predominantly homogeneous nucleation observed for the (2 × 1) islands. Only a few c(6 × 2) islands were found on the terraces, e.g. one in the lower right corner of Fig. 4.16. In the center of this island, part of the topmost Cu layer is removed. Similar holes were found to be typical for isolated c(6 × 2) islands. They indicate mass transport during the (2 × 1) → c(6 × 2) transformation due to a change in the two-dimensional density of surface Cu atoms. A quantitative analysis [4.95] revealed $\Theta_{Cu} = 5/6$ for the c(6 × 2) structure, which is different from the Cu densities in the (2 × 1)O phase and in the clean surface, $\Theta_{Cu} = 0.5$ and $\Theta_{Cu} = 1.0$, respectively. This value for the Cu density and the pronounced two-dimensional corrugation of the c(6 × 2) phase both point to a structure of that phase which is not a simple analog of the (2 × 1) reconstruction. The change in density also provides an explanation for the predominantly heterogeneous nucleation at step edges. Similar to the (2 × 1) formation, Cu atoms are most easily removed from steps and move across the surface to form islands of the new phase. However, both steps are slower during formation of the c(6 × 2) than of the (2 × 1) phase. Because of the lower mobility of Cu atoms on the (2 × 1) surface, c(6 × 2) islands form close to the main sources of Cu atoms, hence close to step sites. The lower mobility is also reflected by the low order of the c(6 × 2) areas formed at 300 K.

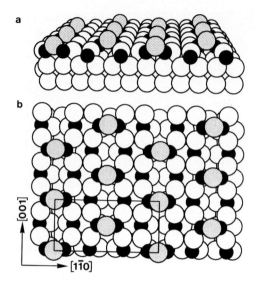

Fig. 4.17. Structure model of the reconstructed Cu(110)–c(6 × 2)O phase; (**a**) side view; (**b**) top view. From [4.95]

The structure model shown in Fig. 4.17 was developed recently from a combined STM/surface X-ray diffraction analysis [4.95]. The proposed structure, which is derived from a (3 × 1) structure, matches the experimental observations very well. In this model the (3 × 1) consists of rows of neighboring –O–Cu–O– strings where every third row is missing. Additional Cu atoms form bridges over the missing-row troughs. These Cu atoms give rise to the dominant, round protrusions in the STM image.

The success and problems in determining the structure of the two O-induced phases on Cu(110) by STM measurements illustrates the possibilities and limitations of this technique for structure determination.

4.3.2 Oxidation Reactions

So far we have discussed STM results for two-dimensional surface reactions. If the coverage of an adsorbate capable of forming strong ionic bonds to the substrate metal atoms is increased beyond a certain level, the substrate/adsorbate reaction usually proceeds into the third dimension, towards the bulk of the metal. Surface oxidation processes represent typical examples for this kind of reaction. The metal ions formed at the surface region are embedded in a lattice of oxygen ions. Usually the density of metal ions in oxides is different from that of the atoms in the substrate, and accordingly the surface topography is expected to change during oxidation. The atomic processes occurring during oxidation reactions are generally not very well understood, and the first STM studies revealed some further information.

Figure 4.18 shows an STM image of a Ni(100) surface which was exposed to 10 L O_2, at room temperature [4.21]. The area displays a sequence of terraces of 100 Å to 200 Å width (falling level from left to right), which are separated by

Fig. 4.18. NiO nuclei (small white dots) and larger oxide islands (e.g. at (*o*)) on Ni(100)(1500 Å × 1200 Å). From [4.21]

parallel, monoatomic steps. Two additional steps (v) originating in screw dislocations (*s*), intersect these terraces. The LEED pattern indicated that most parts of the surface were covered by the c(2 × 2) structure of chemisorbed oxygen atoms, which are not resolved in Fig. 4.18. From higher resolution images, such as those of Fig. 4.2 and Fig. 4.11a, we know that O adatoms on Ni(100) show up as indentations in STM images. Hence, the small white spots in Fig. 4.18 cannot be attributed to O_{ad} atoms. Instead they were interpreted as nuclei of NiO [4.21]. Accordingly, the larger white features of about 50 Å diameter, e.g. at (o) in Fig. 4.18, were associated with larger NiO islands. Although there is a large number of oxide nuclei on the flat terraces, there seems to be a preference for nucleation along step edges. In particular, all the larger islands are centered at step edges. This indicates that either the growth of NiO at step edges is faster or the nucleation of NiO occurs at lower coverages at these places. Pronounced effects of structural imperfections on the rate of oxygen uptake and oxidation were also observed in other studies [4.27]. This is compatible with a simple picture in which the penetration of oxygen atoms into the Ni substrate and the subsequent structural rearrangement of Ni atoms involve activated steps, which have a lower barrier at steps and defect sites. Similar observations were made by STM for the nucleation of aluminum oxide on an Al(111) surface [4.102].

The further development of the oxide layer and ongoing modifications of the surface topography can also be observed, provided the conductivity of the oxide is sufficient for passing the tunnel current. In the case of Ni(100), this was possible for a range of exposures from 5 L to 10000 L. Following an exposure to 5000 L O_2, after which the Ni(100) surface is completely covered by NiO, the STM images showed a strong roughening with a corrugation amplitude of about 50 Å [4.21]. This contrasts the conception of a smooth, thin oxide layer at saturation, which had been developed from kinetic modeling of the oxygen uptake [4.27]. Because of the lower resolution in images of the heavily oxidized Ni surface, it could not be determined from STM whether the NiO films grows epitaxially or disordered. A certain degree of order had been deduced from former LEED experiments [4.27]. A surprisingly regular structure was reported from STM measurements on an oxidized Cr(110) surface [4.103]. The oxide film had been prepared by exposing a cleaved chromium sample to air. The images, which were recorded under ambient pressure, were interpreted as showing an ordered chromium sesquioxide (Cr_2O_3) layer [4.103].

Despite the many problems of in situ studies of oxidation processes, particularly due to instabilities of the tip caused by extended oxygen exposure, STM studies provide a good chance to get more insight into the complex mechanism of these processes.

4.4 Epitaxial Growth of Metals on Metal Substrates

Adsorption of metallic adlayers and subsequent growth of thin metal films represent a particularly tempting task for STM studies. A number of processes, which in the end determine growth mode and topography of the resulting film, have been accessible so far only by indirect means. These processes include the two-dimensional (2D) condensation of adatoms into islands and their subsequent growth, the transition from 2D to three-dimensional (3D) growth and finally, the continuing 3D growth of the film. The different growth modes predicted theoretically [4.104] were verified experimentally from the coverage dependence of spectroscopic intensities [4.105–108] or from the intensity/width of diffraction beams at different coverages [4.109, 110]. Thermal desorption spectroscopy (TDS), reflecting the population of different adsorption states at desorption temperature, was used to gain information on the population of different layers [4.111, 112]. Based on the results of the first STM studies on the structure and growth of Au, Ag and Ni films on Au(111) [4.113–116], of Cu or Au films on Ru(0001) [4.102, 117–120] and on Ag growth on Ni(100) [4.121], we will demonstrate the type of information to be expected from these studies. It will be shown that STM observations give direct information on the structure and topography of the surface, on an atomic scale, and that the nucleation and growth processes described above can be studied and understood to a large extent from systematic STM observations following different procedures for metal deposition and subsequent annealing.

The mobility of metal adatoms deposited at 300 K on metal substrates must be rather high, since individual adatoms have not been observed in room temperature STM studies. Instead, the adatoms were found to condense into islands, which are either nucleated inmid of flat terraces or attached to step edges or other surface defects. These results agree well with existing FIM data, which equally point to a rather high mobility of individual metal adatoms or small 2D clusters of adatoms at 300 K [4.122, 123].

The formation of 2D islands, which represents the first stage of agglomeration, can be described in terms of a nucleation and growth process [4.124]. Following common terminology, the formation of islands inmid of flat terraces can be described as homogeneous nucleation, whereas condensation at step edges or nucleation at surface heterogeneities correspond to heterogeneous nucleation. The existing STM data indicate that in all systems investigated so far metal adatoms condense at the lower terrace side of step edges – on ledge or kink sites. In a few cases nucleation at other types of surface inhomogeneities was also reported. Following room temperature deposition of Ni on a reconstructed Au(111) substrate, small Ni islands were found to nucleate at the bending points of the "herringbone" reconstruction of the underlying, reconstructed Au(111) substrate [4.116]. A rectangular, ordered array of Ni islands is formed with lattice spacings of 73 Å and 140 Å, respectively (Fig. 4.19), corresponding to the unit cell of the reconstruction superlattice. Interestingly, similar findings were reported from an earlier transmission electron microscopy (TEM) study on the nucleation of Cu on a Au(111) substrate [4.125]. At similar deposition conditions Au islands on Ru(0001) showed a high preference to nucleate at points where vertical screw dislocations emerge at the surface. Even at very low coverages, all these points were decorated by isotropic adatom islands, which is shown in the images in Fig. 4.20 [4.120].

Homogeneous nucleation of 2D islands inmid of flat terraces was found as a competing mechanism to heterogeneous nucleation at step edges, which can occur only when the terraces are sufficiently wide [4.120]. The density of islands is often determined by kinetic restrictions. Close to thermodynamic equilibrium only few islands of a rather compact shape should exist. Experimentally, however this quantity is often found to strongly depend on the deposition conditions, i.e. mainly on the flux of metal atoms and the substrate temperature. Room temperature deposition of Au on a Ru(0001) substrate represents a good example. A detailed study showed that above a critical coverage the island density remains constant [4.120]. Further deposition leads only to an increase in island size. Similar behavior would be expected from a 2D nucleation-and-growth process, where the metal adatoms migrate freely over the substrate surface until they form a stable nucleus or condense at an existing nucleus/island. Within a certain area around each existing island, the probability for homogeneous nucleation of additional islands is very low. The width

Fig. 4.19. STM image of Ni island arrays on Au(111). Three atomically flat Au terraces are seen, separated by steps of single-atom height. Small light dots on each terrace are monolayer Ni islands. Ni coverage is 0.11 ± 0.03 ML deposited at 0.4 ML/min. The central terrace has a boundary between two island "hyperdomains". From [4.116]

of this area is given by the effective mean free path of the mobile adatoms, i.e. the distance the adatoms can travel before they are trapped in a nucleation or condensation process. All adatoms impinging in this area can reach the perimeter of the next island and condense there [4.120]. The same effect inhibits island formation in substrate areas near step edges, which was experimentally verified in the same study. Examples for such behavior are seen in the images in Fig. 4.20.

Kinetic effects can also govern the 2D growth of the nuclei. Kinetic limitations often cause the formation of rather irregular island shapes. Highly dendritic growth was demonstrated for Au deposits on Ru(0001), as shown in the sequence of images in Fig. 4.20 [4.120]. Growth of similar structures was simulated by *Witten* and *Sander* in the framework of the diffusion-limited aggregation model [4.126]. In this model they assume the adatoms to move in a random walk over the surface and to remain adsorbed at the perimeter of an island wherever they contact it first ("hit and stick"). Diffusion along the island perimeter is prohibited in the most simple version of this model. They predicted a constant, fractal dimension of these islands over a wide coverage range and a power law decay of the adatom density in the island with increasing distance from the center of the island. Both of these predictions were verified experimentally, as described in more detail in [4.120].

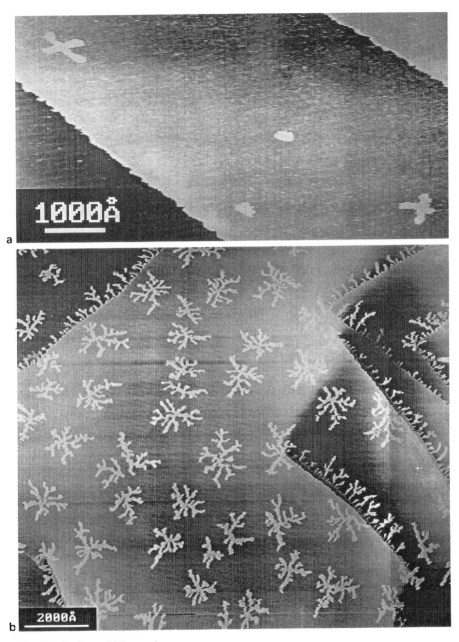

Fig. 4.20a,b. See p. 75 for caption

Fig. 4.20c,d. See opposite page for caption

These dendritic-shaped islands are thermally instable and collapse into more compact entities upon annealing. It is interesting to note that heating experiments confirmed the existence of two distinctly different temperature regimes for adatom diffusion. At lower annealing temperatures, adatoms can only move along the island perimeters. In this case the shape of the island is changed, but there is no appreciable material transport between islands so the mass per island remains constant. At higher temperatures, adatoms can also dissolve from the island perimeter and islands can either grow at the expense of others or can be completely dissolved. Upon deposition at higher temperatures, dendritic growth does not occur and compact islands are formed immediately [4.120].

The transition from 2D growth to 3D growth and the subsequent 3D growth have been extensively investigated both theoretically [4.104, 127] and experimentally [4.105, 106]. Different models were proposed based on thermodynamic considerations, namely layer-by-layer growth (Frank-van der Merwe), layer-plus-cluster growth (Stranski-Krastanov) and immediate growth of 3D clusters (Volmer-Weber) [4.104, 105]. In most cases, however, kinetic effects have considerable influence or even solely determine the film growth [4.128]. For example, thermal desorption spectra show that the first layer of Au atoms on a Ru(0001) substrate is thermodynamically more stable than subsequent layers. Nevertheless, for room temperature deposition (flux ca. two monolayers (ML) per min) of Au on Ru(0001) it was shown that second layer nucleation starts long before the first layer is filled, at a total coverage of ca. 0.8 ML [4.119]. For Au on Au(111) second layer population was found already at the rather low coverage of 0.34 ML upon deposition at room temperature. *Lang* et al. followed the growth process of the same system up to higher coverages [4.114]. At nominal coverages between 1.0 and 6.0 ML they observed Au islands populating several different layers of the film. An even more extreme example was found for room temperature deposition of Cu on Ru(0001) [4.117]. In the image in Fig. 4.21, which shows a surface covered by nominally ~5 ML, at least 10 subsequent layers are populated simultaneously. Pyramidal, quasi-3D structures were formed from stacks of the successively smaller Cu islands.

The STM studies mentioned above directly show how the transition from 2D to 3D growth occurs by nucleation of small second-layer islands on top of first-layer islands. Following the ideas on homogeneous island nucleation derived above, nucleation of second-layer islands, on top of first-layer islands, can occur as soon as the diameter of the latter islands exceeds the effective mean free path of second layer adatoms. For heteroepitaxial systems, this quantity

Fig. 4.20a–d. Sequence of STM topographs (ca. 1 μm × 1 μm) recorded on a Ru(0001) surface covered with submonolayer Au films which were deposited at room temperature (2 ML/min). Islands are formed inmid of large terraces, while in areas along steps the Au atoms aggregate at the step edges. These images show the dendritic 2D-growth of the islands with increasing coverage. At higher coverages the arms of the islands begin to thicken due to adsorption on top and between the arms. (**a**) $\Theta \approx 0.03$ ML, (**b**) $\Theta \approx 0.15$ ML, (**c**) $\Theta \approx 0.40$ ML, (**d**) $\Theta \approx 0.70$ ML. From [4.120]

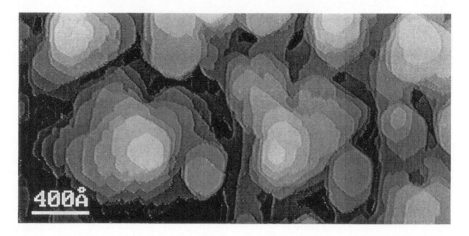

Fig. 4.21. STM topograph (3000 Å × 1450 Å) of a thicker Cu film ($\Theta_{Cu} \sim 5$ ML) deposited at 300 K on a Ru(0001) substrate. Pyramidal structures are formed by increasingly smaller islands in subsequent layers, ≥ 10 layers are populated simultaneously. From [4.118]

may be significantly different from that of first-layer adatoms, reflecting the different adatom mobilities on the respective substrates. A reduced mobility in the higher layer will lead to a lower value for the mean free path of the adatoms. Hence nuclei can be formed on top of first-layer islands before these have grown together and can continue with 2D growth. This leads to the observed quasi-3D growth. As pointed out by *Kunkel* et al. [4.129], this mechanism can explain deviations from layer-by-layer growth even for homoepitaxial systems. In their model the density of adatoms on first-layer islands, and hence their mean free path, can differ from that on the bulk substrate due to repulsive interactions between the descending step edges at the perimeter of the first-layer islands and second-layer adatoms [4.129]. These energy barriers hinder second-layer adatoms from crossing the step at the island perimeters, which is necessary for them to reach the lower terraces where they can finally condense at the edges of the islands. Such repulsions were in fact derived from FIM observations, e.g. for W atoms on a W(110) substrate [4.130].

In addition to characterizing the growth behavior, STM studies provide a direct means to determine the periodic and defect structure of the film independently for each layer. The lattice structure in the interface region must transform from that of the substrate to that of the bulk overlayer material. The misfit between these lattices leads to a significant lattice strain in the interface region which, following early predictions by *Frank* and *van der Merwe*, is relieved by the formation of misfit dislocations in the film [4.131]. The existence of these misfit dislocations was first verified and intensely studied by transmission electron microscopy [4.125]. In most cases, however, it was not possible to unravel the exact nature of the structural transition between the respective lattices.

An STM study on Cu/Ru(0001) [4.117–119] confirmed earlier LEED observations [4.111, 132, 133] that in this system a structural transition takes place between a pseudomorphic first layer and a contracted second layer. In the first layer the more tightly bound Cu adatoms are forced into the geometry of the substrate lattice, leading to an isotropic expansion of 5.5% as compared to the Cu(III) lattice and thus to significant strain in that layer. Though the hcp adsorption site is more favorable, as concluded from a structural analysis by LEED [4.134], STM observations revealed the formation of triangular fcc

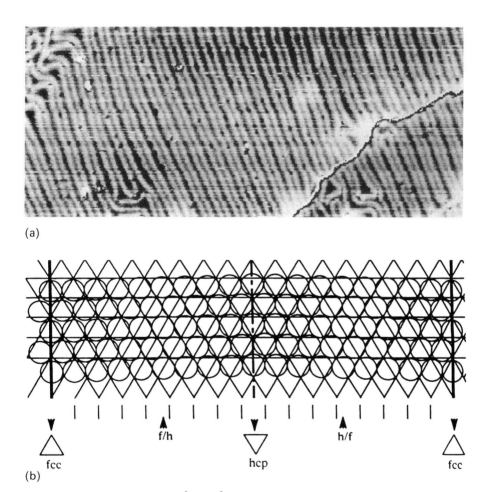

(a)

(b)

Fig. 4.22. (a) STM topograph (1200 Å × 470 Å) of a Cu-covered Ru(0001) surface after annealing to 1000 K ($\Theta_{Cu} \approx 2$ ML), exhibiting large, well-ordered domains of the reconstructed Cu bilayer phase on two consecutive terraces. **(b)** Topview model of the atomic structure of the unidirectionally contracted second Cu layer. The hexagonal grid marks the position of the first, pseudomorphic layer Cu atoms. The second layer Cu atoms are represented by large circles. The alternating occupation of hcp and fcc threefold-hollow sites requires lateral displacements in [120] direction. From [4.119]

domains as a minority species upon deposition at room temperature. STM images of the second layer resolve a characteristic corrugation consisting of parallel pairs of lines in the [120] direction, ca. 46 Å apart (Fig. 4.22a). These images very much resemble those of the reconstructed Au(111) surface [4.135, 136]. They can be explained in terms of a similar structural model, namely by a sequence of hcp and fcc stacking domains separated by partial dislocations [4.135–138]. It is evident from the model in Fig. 4.22b that the Cu atoms in the dislocation zones are located on quasi-bridge sites and should appear more prominently than the adatoms on threefold-hollow sites in the hcp and fcc regions. Hence in the second Cu layer the lattice strain is reduced by a unidirectional contraction of the lattice along the close-packed [100] direction of the hexagonal substrate. This result differs from earlier predictions of an isotropic contraction [4.132, 133] and leads to a "striped structure" with parallel partial misfit dislocations along [120]. In subsequent layers the lattice must contract also in the other directions, e.g. by formation of additional dislocations, until the final, isotropically contracted lattice of the epitaxially grown Cu(111) surface is reached.

In contrast to the above example, where the "smaller" Cu atoms ultimately lead to a lattice contraction, deposition of Au on the same substrate should lead to an expansion of the lattice because of its larger lattice parameters. LEED observations indeed reveal a reconstruction in the first layer [4.139–141], whose structure was explained in a very similar model of a striped structure, namely a unidirectional expansion with parallel partial dislocation lines between sequential fcc and hcp stacking domains [4.139–141]. STM observations of a striped corrugation in the first Au layer, similar to that observed in the second layer for Cu deposition, confirm this assignment [4.118]. For both of the reconstructed adlayers STM images reveal the presence of a significant number of domain boundaries upon room temperature deposition. In particular for the Cu bilayer, well-ordered structures are formed only upon annealing or for deposition at higher temperatures [4.119]. Hence the ordering process is thermally activated. At lower temperatures the contraction/expansion process occurs on a local scale without achieving a long-range order.

In the examples presented above the structural transition in the interface region occurs either directly at the interface or between subsequent layers of the deposited material. But this is not necessarily the thermodynamically most stable structure for the interface region. In general structural transitions can also occur in the interface region of the substrate. In most cases, however, these transitions will occur only after sufficiently thick films are deposited and after additional thermal treatment in order to achieve the thermodynamically stable structure of the interface region. These deeper lying dislocations are not accessible for STM measurements. In a few cases, however, structural transitions in the substrate can be induced already by very thin films. One example for such behavior was found for Ag deposits on Au(111). It was shown by TEM studies [4.125] and recently confirmed by STM that the reconstruction of the clean Au(111) surface is removed by the presence of the Ag adlayer [4.113, 115].

These results give a first impression of the extent to which STM studies can resolve structural effects and provide insight into the different processes involved in thin film growth on an atomic scale.

4.5 Conclusions

In this chapter we have reviewed the current state of STM investigations on the interaction of adsorbates with metal surfaces. The representation of individual adsorbate species and the resolution of closed adlayers in STM images were discussed and compared with predictions from theory. The characterization of the local adsorption complex by STM was shown to give access to details of the chemical bond between substrate and adsorbate. High-resolution images can reveal details of complex, periodic structures and can thus assist other methods for structure determination by reducing the number of possible structure models. STM measurements on the local order, distribution and mobility of adsorbates allow conclusions to be drawn on the effects of adsorbate–adsorbate interactions and, most importantly, on the role of surface defects on these properties. The use of STM studies for revealing mechanistic details of surface reactions was discussed, based on different examples. As demonstrated, completely new mechanistic concepts could already be derived in several cases. Finally, we presented results of the first STM studies on the morphology and growth process, and on the atomic structure of thin metallic films on metal substrates. They allow a direct test of the different models developed to describe film growth and strain relaxation within the film.

Acknowledgments. We gratefully acknowledge F. Besenbacher, D.D. Chambliss, D.M. Eigler, Y. Kuk, N.D. Lang, H. Neddermeyer and M. Salmeron, who kindly provided illustrations of their work for inclusion in this chapter.

References

4.1   E. Bauer: In *Chemistry and Physics of Solid Surfaces VIII*, ed. by R. Vanselow, R. Howe (Springer, Berlin, Heidelberg 1990) p. 267
4.2   A.J. Melmed: In *Chemistry and Physics of Solid Surfaces VI*, ed. by R. Vanselow, R. Howe (Springer, Berlin, Heidelberg 1986) p. 325
4.3   D.M. Eigler, E.K. Schweizer: Nature **344**, 524 (1990)
4.4   N.D. Lang, A.R. Williams: Phys. Rev. **B25**, 2940 (1982)
4.5   E. Kopatzki, R.J. Behm: Surface Sci. **245**, 255 (1991)
4.6   W. Oed, H. Lindner, U. Starke, K. Heinz, K. Müller, J.B. Pendry: Surface Sci. **224**, 179 (1989)
4.7   L. Wenzel, D. Arvanitis, W. Daum, H.H. Rotermund, J. Stöhr, K. Baberschke, H. Ibach: Phys. Rev. B **36**, 7689 (1987)
4.8   J. Tersoff, D.R. Hamann: Phys. Rev. Lett. **50**, 1998 (1983)
4.9   J. Tersoff, D.R. Hamann: Phys. Rev. B **31**, 805 (1985)
4.10 R.J. Behm, W. Hösler: In *Chemistry and Physics of Solid Surfaces VI*, ed. by R. Vanselow, R. Howe (Springer, Berlin, Heidelberg 1986) p. 361
4.11 J. Bardeen: Phys. Rev. Lett. **6**, 57 (1961)

4.12 N.D. Lang: Phys. Rev. Lett. **55,** 230 (1985)
4.13 N.D. Lang: Phys. Rev. Lett. **56,** 1164 (1986)
4.14 N.D. Lang: IBM J. Res. Develop. **30,** 374 (1986)
4.15 G. Doyen, D. Drakova, E. Kopatzki, R.J. Behm: J. Vac. Sci. Technol. A **6,** 327 (1988)
4.16 E. Kopatzki, G. Doyen, D. Drakova, R.J. Behm: J. Microsc. **152,** 687 (1988)
4.17 H.H. Rotermund, K. Jacobi: Surface Sci. **126,** 32 (1983)
4.18 K. Horn, K.H. Frank, J.A. Wilder, B. Reihl: Phys. Rev. Lett. **57,** 1064 (1986)
4.19 K. Wandelt, W. Jacob, N. Memmel, V. Dose: Phys. Rev. Lett. **57,** 1643 (1986)
4.20 N.D. Lang: Phys. Rev. Lett. **46,** 842 (1981)
4.21 G. Wilhelmi, A. Brodde, D. Badt, H. Wengelnik, H. Neddermeyer: In *The Structure of Surfaces III,* ed. by S.Y. Tong, M.A. Van Hove, X. Xide (Springer, Berlin, Heidelberg 1991) p. 448
4.22 J. Wintterlin, H. Brune, H. Höfer, R.J. Behm: Appl. Phys. A **47,** 99 (1988)
4.23 C. Günther, R.J. Behm: To be published
4.24 D. Norman, S. Brennan, R. Jaeger, J. Stöhr: Surface Sci. **105,** L297 (1981)
4.25 M. Lindroos, H. Pfnür, G. Held, D. Menzel: Surface Sci. **222,** 451 (1989)
4.26 W. Eberhardt, F.J. Himpsel: Phys. Rev. Lett. **42,** 1375 (1979)
4.27 C.R. Brundle, J. Broughton: In *The Chemical Physics of Solid Surfaces and Heterogeneous Catalysis,* Vol. 3A, ed. by D.A. King, D.P. Woodruff (Elsevier, Amsterdam, 1990)
4.28 J.C. Fuggle, T.E. Madey, M. Steinkilberg, D. Menzel: Surface Sci. **52,** 521 (1975)
4.29 N.D. Lang: Comments Condens. Matter Phys. **14,** 253 (1989)
4.30 N.D. Lang: Surface Sci. **127,** L118 (1983)
4.31 H. Scheidt, M. Glöbl, V. Dose: Surface Sci. **123,** L728 (1982)
4.32 H. Brune, J. Wintterlin, G. Ertl, R.J. Behm: To be published
4.33 H. Brune, J. Wintterlin, G. Ertl, R.J. Behm: Europhys. Lett. **13,** 123 (1990)
4.34 D.J. Coulman, J. Wintterlin, R.J. Behm, G. Ertl: Phys. Rev. Lett. **64,** 1761 (1990)
4.35 J. Wintterlin, J. Wiechers, Th. Gritsch, H. Höfer, R.J. Behm: J. Microsc. **152,** 423 (1988)
4.36 J. Wintterlin, J. Wiechers, H. Brune, T. Gritsch, H. Höfer, R.J. Behm: Phys. Rev. Lett. **62,** 59 (1989)
4.37 Y. Kuk, P.J. Silverman, F.M. Chua: J. Microsc. **152,** 449 (1988)
4.38 P.H. Lippel, R.J. Wilson, M.D. Miller, Ch. Wöll, S. Chiang: Phys. Rev. Lett. **62,** 171 (1989)
4.39 J. Tersoff: Phys. Rev. B **39,** 1052 (1989)
4.40 H. Ohtani, R.J. Wilson, S. Chiang, C.M. Mate: Phys. Rev. Lett. **60,** 2398 (1988)
4.41 D.F. Ogletree, C. Ocal, B. Marchon, G.A. Somorjai, M. Salmeron, T. Beebe, W. Siekhaus: J. Vac. Sci. Technol. A **8,** 297 (1990)
4.42 J.M. MacLaren, J.B. Pendry, P.J. Rous, D.K. Saldin, G.A. Somorjai, M.A. Van Hove, D.D. Vvedensky: *Surface Crystallographic Information Service* (Reidel, Dordrecht, 1987)
4.43 D.G. Kelly, A.J. Gellman, M. Salmeron, G.A. Somorjai, V. Maurice, M. Huber, J. Oudar: Surface Sci. **204,** 1 (1988)
4.44 B. Marchon, D.F. Ogletree, M.E. Bussell, G.A. Somorjai, M. Salmeron, W. Siekhaus: J. Microsc. **152,** 427 (1988)
4.45 B. Marchon, P. Bernhardt, M.E. Bussell, G.A. Somorjai, M. Salmeron, W. Sieckhaus: Phys. Rev. Lett. **60,** 1166 (1988)
4.46 S. Rousset, S. Gauthier, O. Siboulet, W. Sacks, M. Belin, J. Klein: Phys. Rev. Lett. **63,** 1265 (1989)
4.47 S. Rousset, S. Gauthier, O. Siboulet, W. Sacks, M. Belin, J. Klein: J. Vac. Sci. Technol. A **8,** 302 (1990)
4.48 B.C. Schardt, S.-L. Yau, F. Rinaldi: Science **243,** 1050 (1989)
4.49 G.F.A. van de Walle, H. van Kempen, P. Wyder, C.J. Flipse: Surface Sci. **181,** 27 (1987)
4.50 T. Gritsch, G. Ertl, R.J. Behm: To be published
4.51 W. Hösler, R.J. Behm, E. Ritter: IBM J. Res. Develop. **30,** 403 (1986)
4.52 E. Ritter, R.J. Behm, G. Pötschke, J. Wintterlin: Surface Sci. **181,** 403 (1987)
4.53 W. Hösler, E. Ritter, R.J. Behm: Ber. Bunsenges. Phys. Chem. **90,** 205 (1986)
4.54 S. Chiang, R.J. Wilson, C.M. Mate, H. Ohtani: J. Microsc. **152,** 567 (1988)
4.55 P. Hofmann, S.R. Bare, D.A. King: Surface Sci. **117,** 245 (1982)

4.56 A. Brodde, St. Tosch, H. Neddermeyer: J. Microsc. **152,** 441 (1988)
4.57 G. Binnig, K.H. Frank, H. Fuchs, N. García, B. Reihl, H. Rohrer, F. Salvan, A.R. Williams: Phys. Rev. Lett. **55,** 991 (1985)
4.58 C. Ettl, R.J. Behm: Unpublished data
4.59 H.J. Freund, J. Ragozik, V. Dose, M. Neumann: Surface Sci. **175,** 651 (1986)
4.60 H. Kuhlenbeck, M. Neumann, H.J. Freund: Surface Sci. **173,** 94 (1986)
4.61 R.M. Feenstra, J.A. Stroscio, A.P. Fein: Surface Sci. **181,** 295 (1987)
4.62 N.D. Lang: Phys. Rev. B **34,** 5947 (1986)
4.63 F.M. Chua, Y. Kuk, P.J. Silverman: Phys. Rev. Lett. **63,** 386 (1989)
4.64 W. Jacob, V. Dose, A. Goldmann: Appl. Phys. A **41,** 145 (1986)
4.65 R.A. DiDio, D.M. Zehner, E.W. Plummer: J. Vac. Sci. Technol. A2, 852 (1984)
4.66 N. Lang: Phys. Rev. Lett. **58,** 45 (1987)
4.67 W. Sesselmann, H. Conrad, G. Ertl, J. Küppers, B. Woratschek, H. Haberland: Phys. Rev. Lett. **50,** 446 (1983)
4.68 Y. Kuk, M.F. Jarrold, P.J. Silverman, J.E. Bower, W.L. Brown: Phys. Rev. B **39,** 11168 (1989)
4.69 P.K. Hansma (Ed.): *Tunneling Spectroscopy* (Plenum, New York 1982)
4.70 D.M. Eigler: Private communication
4.71 B.N.J. Persson, J. Demuth: Solid State Commun. **57,** 769 (1986)
4.72 H.S. Taylor: Proc. Roy. Soc. (Lond.) A **108,** 105 (1925)
4.73 E.K. Schweizer, D.M. Eigler: unpubl. data
4.74 H. Brune, J. Wintterlin, G. Ertl, R.J. Behm: To be published
4.75 S.M. Davies, G.A. Somorjai: In *The Chemical Physics of Solid Surfaces and Heterogeneous Catalysis*, Vol. 4, ed. by D.A. King, D.P. Woodruff (Elsevier, Amsterdam 1982) p. 217
4.76 G. Binnig, H. Fuchs, E. Stoll: Surface Sci. **169,** L295 (1986)
4.77 F. Jensen, F. Besenbacher, E. Lægsgaard, I. Stensgaard: Phys. Rev. B **41,** 10233 (1990)
4.78 Y. Kuk, F.M. Chua, P.J. Silverman, J.A. Meyer: Phys. Rev. B **41,** 12393 (1990)
4.79 J. Wintterlin, R. Schuster, D.J. Coulman, G. Ertl, R.J. Behm: J. Vac. Sci. Technol. B **9,** 902 (1991)
4.80 R.C. Jaklevic, L. Elie: Phys. Rev. Lett. **60,** 120 (1988)
4.81 R. Emch, J. Nogami, M.M. Dovek, C.A. Lang, C.F. Quate: J. Microsc. **152,** 129 (1988)
4.82 M. Henzler: In *Reflection High-Energy Electron Diffraction and Reflection Electron Imaging of Surfaces*, ed. by P.K. Larsen, P.J. Dobson (Plenum, New York 1988) p. 193
4.83 K.M. Ho, K.P. Bohnen: Phys. Rev. Lett. **59,** 1833 (1987)
4.84 B.W. Dodson: Phys. Rev. Lett. **60,** 2288 (1988)
4.85 G. Binnig, H. Rohrer, Ch. Gerber, E. Weibel: Surface Sci. **131,** L379 (1983)
4.86 A.M. Baró, G. Binnig, H. Rohrer, Ch. Gerber, E. Stoll, A. Baratoff, F. Salvan: Phys. Rev. Lett. **52,** 1304 (1984)
4.87 R.J. Behm, W. Hösler, E. Ritter, G. Binnig: J. Vac. Sci. Technol. A **4,** 1330 (1986)
4.88 R.J. Behm, W. Hösler, E. Ritter, G. Binnig: Phys. Rev. Lett. **56,** 228 (1986)
4.89 Y. Kuk, P.J. Silverman, H.Q. Nguyen: Phys. Rev. Lett. **59,** 1452 (1987)
4.90 Y. Kuk, P.J. Silverman, H.Q. Nguyen: J. Vac. Sci. Technol. A **6,** 524 (1988)
4.91 Y. Kuk, P.J. Silverman, H.Q. Nguyen: J. Vac. Sci. Technol. A **6,** 567 (1988)
4.92 T. Gritsch, D. Coulman, R.J. Behm, G. Ertl: Appl. Phys. A **49,** 403 (1989)
4.93 T. Gritsch, D. Coulman, R.J. Behm, G. Ertl: Phys. Rev. Lett. **63,** 1086 (1989)
4.94 F.M. Chua, Y. Kuk, P.J. Silverman: J. Vac. Sci. Technol. A **8,** 305 (1990)
4.95 R. Feidenhans'l, F. Grey, M. Nielsen, F. Besenbacher, F. Jensen, E. Lægsgaard, I. Stensgaard, K.W. Jacobsen, J.K. Nørskov, R.L. Johnson: Phys. Rev. Lett. **65,** 2027 (1990)
4.96 D. Coulman, J. Wintterlin, J.V. Barth, G. Ertl, R.J. Behm: Surface Sci. **240,** 151 (1990)
4.97 F. Jensen, F. Besenbacher, E. Lægsgaard, I. Stensgaard: Phys. Rev. B **42,** 9206 (1990)
4.98 R. Schuster, J.V. Barth, G. Ertl, R.J. Behm: Surface Sci. **247,** L229 (1991)
4.99 R.P.N. Bronckers, A.G.J. de Wit: Surface Sci. **112,** 133 (1981)
4.100 R. Feidenhans'l, I. Stensgaard: Surface Sci. **133,** 453 (1983)
4.101 J.M. Mundenar, A.P. Baddorf, E.W. Plummer, L.G. Sneddon, R.A. DiDio, D.M. Zehner: Surface Sci. **188,** 15 (1987)

4.102 R.J. Behm: In *Scanning tunneling microscopy and related methods*, ed. by R.J. Behm, N. García, H. Rohrer (Kluwer, Dordrecht 1990) p. 173
4.103 N.M.D. Brown, H.-X. You: Surface Sci. **233**, 317 (1990)
4.104 E. Bauer: Z. Krist. **110**, 372 (1958)
4.105 E. Bauer: Appl. Surface Sci. **11/12**, 479 (1982)
4.106 E. Bauer: In *The Chemical Physics of Solid Surfaces and Heterogeneous Catalysis*, Vol. 3B, ed. by D.A. King, D.P. Woodruff (Elsevier, Amsterdam 1984) p. 1
4.107 G.E. Rhead: J. Vac. Sci. Technol. **13**, 603 (1976)
4.108 A. Jablonski, S. Eder, K. Wandelt: Applic. Surface Sci. **22/23**, 309 (1985)
4.109 C. Lilienkamp, C. Koziol, E. Bauer: In *Reflection High Energy Electron Diffraction and Reflection Electron Imaging of Surfaces*, ed. by P.K. Larsen, P.J. Dobson (Plenum, New York 1988) p. 489
4.110 M. Henzler: Appl. Surface Sci. **11/12**, 450 (1982)
4.111 K. Christmann, G. Ertl, H. Shimizu: Thin Solid Films **57**, 247 (1979); J. Catal. **61**, 397 (1980)
4.112 J.W. Niemandtsverdriet, P. Dolle, K. Markert, K. Wandelt: J. Vac. Sci. Technol. A **5**, 2849 (1987)
4.113 M.M. Dovek, C.A. Lang, J. Nogami, C.F. Quate: Phys. Rev. B **40**, 11973 (1989)
4.114 C.A. Lang, M.M. Dovek, J. Nogami, C.F. Quate: Surface Sci. **224**, L947 (1989)
4.115 D.D. Chambliss, R.J. Wilson: J. Vac. Sci. Technol. B **9**, 928 (1991)
4.116 D.D. Chambliss, R.J. Wilson, S. Chiang: Phys. Rev. Lett. **66**, 1721 (1991); J. Vac. Sci. Technol. B **9**, 933 (1991)
4.117 G. Pötschke: Ph.D. thesis, University of Munich (1990)
4.118 G. Pötschke, J. Schröder, C. Günther, R.Q. Hwang, R.J. Behm: Surface Sci. **251/252**, 592 (1991)
4.119 G. Pötschke, R.J. Behm: Phys. Rev. B **44**, 1442 (1991)
4.120 R.Q. Hwang, J. Schröder, C. Günther, R.J. Behm: Phys. Rev. Lett. **67**, 3279 (1991)
4.121 T. Brodde, G. Wilhelmi, G. Badt, H. Wengelnik, H. Neddermeyer: J. Vac. Sci. Technol. B **9**, 920 (1991)
4.122 G. Ehrlich: In *Proc. 9th Int. Vac. Conf. and 5th Intern. Conf. Solid Surfaces* (Madrid, 1983) p. 3
4.123 G. Ehrlich: In *Chemistry and Physics of Solid Surfaces V*, ed. by R. Vanselow, R. Howe (Springer, Berlin, Heidelberg 1984) p. 283
4.124 R. Kern: In *Interfacial Aspects of Phase Transitions*, ed. by B. Mutaftschiev (Reidel, Dordrecht 1982) p. 287
4.125 K. Yagi, K. Tobayashi, Y. Tanishiro, K. Takanayagi: Thin Solid Films **126**, 95 (1985) and references therein
4.126 T.A. Witten, L.M. Sander: Phys. Rev. B **27**, 5686 (1983)
4.127 S. Stoyanov, I. Markov: Surface Sci. **116**, 313 (1982) and references therein
4.128 D. Kaishev: J. Crystal Growth **40**, 29 (1977); Thin Solid Films **55**, 399 (1978)
4.129 R. Kunkel, B. Poelsema, L. Verheij, G. Comsa: Phys. Rev. Lett. **65**, 773 (1990)
4.130 G. Ehrlich, F.G. Hudda: J. Chem. Phys. **44**, 1030 (1966)
4.131 F.C. Frank, J.H. van der Merwe: Proc. R. Soc. **198**, 205 (1949)
4.132 J.E. Houston, C.H.F. Peden, D.S. Blair, D.W. Goodman: Surface Sci. **167**, 427 (1986)
4.133 C. Park, E. Bauer, H. Poppa: Surface Sci. **187**, 86 (1987)
4.134 H. Davies: unpubl. data
4.135 C. Wöll, S. Chiang, R.J. Wilson, P.H. Lippel: Phys. Rev. B **39**, 7988 (1989)
4.136 J.V. Barth, H. Brune, G. Ertl, R.J. Behm: Phys. Rev. B **42**, 9307 (1990)
4.137 Y. Tanishiro, H. Kanamori, K. Takanayagi, K. Yagi, G. Honjo: Surface Sci. **111**, 395 (1981)
4.138 U. Harten, A.M. Lahee, J.P. Toennies, C. Wöll: Phys. Rev. Lett. **54**, 2619 (1985)
4.139 B. Konrad, F.J. Himpsel, W. Steinmann, K. Wandelt: In *Proc. Int. ER-LEED Conf.* (Erlangen, 1985) p. 109
4.140 B. Konrad, D. Rieger, R.D. Schnell, W. Steinmann, K. Wandelt: in *BESSY Jahresber.* (Berlin 1985)
4.141 C. Harendt, K. Christmann, W. Hirschwald: Surface Sci. **165**, 413 (1986)

5. STM on Semiconductors

R.J. Hamers

With 29 Figures

Semiconductors represent the class of compounds most extensively studied using STM, for several reasons. In addition to the great technological importance of semiconductors, they have several unique characteristics which make them particularly attractive candidates for STM studies. The localized nature of the *sp* bonding is particularly attractive for STM imaging, since it provides for large corrugations. Reconstructions are common for semiconductor surfaces, providing a rich variety of interesting topography which is significantly different from the bulk and is not easily determined using conventional surface science tools. Finally, the existence of localized surface states makes measuring and understanding the electronic properties of semiconductors an interesting and challenging scientific problem.

5.1 Experimental Technique

Before reviewing the applications of STM and tunneling spectroscopy to semiconductors, we first review both STM topographic imaging and tunneling microscopy, as applied to semiconductors. In addition to these standard techniques, several modified STM techniques have been developed which provide other unique kinds of information about semiconductors on atomic and nanometer scales. These techniques, including surface photovoltage measurements, tunneling-induced photon emission, and tunneling potentiometry at semiconductor heterostructures, will be discussed at the end.

5.1.1 Topographic Imaging

Measuring the conventional STM topography is virtually always the first experiment performed with a scanning tunneling microscope. On large distance scales, these measurements provide useful information about the overall inclination of the surface, the shape of step edges, and other aspects of the general surface morphology. On a finer scale, interpreting these results can be both extremely interesting and extremely challenging, due once again to the presence of localized surface states associated with the surface atoms.

What does the STM topography measure on a semiconductor surface? At positive sample bias, the net tunneling current arises from electrons which tunnel from the occupied states of the tip into unoccupied states of the sample, so that the contour which the STM tip follows is directly related to the spatial distribution of the occupied electronic states of the sample. At negative sample bias the situation is reversed, and electrons tunnel from occupied states of the sample into unoccupied states of the tip, and the STM tip follows a contour which is related to the spatial distribution of the unoccupied, or empty, electronic states.

For any given lateral position of the tip above the sample (r), the tunneling current (I) is determined by the sample–tip separation (Z), the applied voltage (V), and the electronic structure of the sample and tip which is quantitatively described by their respective density of states ($\varrho(E)$). The Tersoff–Hamann theory of STM [5.1, 2] states that in the low-voltage limit, the STM measures a contour of constant Fermi-level density of states. Unfortunately, on semiconductors the bias voltages used are typically on the order of several volts, so that the assumptions built into the Tersoff–Hamann theory are violated. As a result, we resort to a simpler theory, the WKB theory for planar tunneling. According to this theory, the tunneling current can be expressed as

$$I = \int_0^{eV} \varrho_s(r, E)\varrho_t(r, -eV + E)T(E, eV, r)dE \ , \tag{5.1}$$

where $\varrho_s(r, E)$ and $\varrho_t(r, E)$ are the density of states of the sample and tip at location r and the energy E, measured with respect to their individual Fermi levels. For negative sample bias, $eV < 0$ and for positive sample bias, $eV > 0$. The tunneling transmission probability $T(E, eV, r)$ for electrons with energy E and applied bias voltage V is given by

$$T(E, eV) = \exp\left(-\frac{2Z\sqrt{2m}}{\hbar}\sqrt{\frac{\phi_s + \phi_t}{2} + \frac{eV}{2} - E}\right). \tag{5.2}$$

At constant tunneling current I, the contour followed by the tip is a relatively complicated function of the density of both sample and tip, together with the tunneling transmission probability. However, examination of the transmission probability T shows that if $eV < 0$ (i.e., negative sample bias), the transmission probability is largest for $E = 0$ (corresponding to electrons at the Fermi level of the sample). Similarly, if $eV > 0$ (positive sample bias) the probability is largest for $E = eV$ (corresponding to electrons at the Fermi level of the tip). Thus, we see that the tunneling probability is always largest for electrons at the Fermi level of whichever electrode is negatively biased. The width of the electron distribution depends on the work functions of the materials involved, but for typical work functions of $\simeq 3$–4 eV most of the tunneling current originates from within 300 mV of the Fermi level, but with substantial contributions as much as 1 eV below E_F. Thus, assuming that the density of states of the tip is constant, by choosing the bias voltage properly it is possible to tunnel into discrete *unoccupied* states of the sample. When trying to selectively image *occupied* states of the sample, however, most of the tunneling current will virtually always originate at the state nearest the Fermi energy. On Si(111)–(7 × 7), for example, states arising from the stacking fault occur at energies of 0.2 eV below E_F and 1.4 eV above E_F; constant-current topographs at negative voltages *always* reveal this asymmetry, while images at positive bias only show an asymmetry over a narrow voltage range. This difference arises from the different energy distributions of the tunneling electrons when tunneling from occupied and into unoccupied states of the sample.

5.1.2 Tunneling Spectroscopy

Tunneling spectroscopy provides information complementary to the information obtained in conventional topographic imaging [5.3]. By measuring the detailed dependence of the tunneling current on the applied voltage, it is possible to work backwards through the tunneling equations and to extract a measure of the electronic density of states of the sample. By knowing both the energies and the spatial locations of the electronic states, it is often possible to make direct comparisons with theory.

Tunneling spectroscopy can be accomplished in a number of ways. The main idea of all tunneling spectroscopy experiments is to measure how the tunneling current depends on the applied voltage. The experimental implementation of tunneling spectroscopy can vary depending on the energy range accessed, the amount of spectroscopic detail required, and whether or not high spatial resolution is simultaneously required. A detailed discussion of these various experimental techniques and the interpretation of tunneling spectroscopy data has been presented in a recent review paper [5.3] and will not be presented here.

Most tunneling spectroscopy investigations today are made under conditions of constant separation, which is accomplished by momentarily interrupting the feedback controller and then ramping the applied voltage over the desired interval while simultaneously measuring the tunneling current. If no spatial resolution is required, the method is straightforward [5.4]. In order to be able to correlate the tunneling spectra with the local surface topography (and thus to take full advantage of the high spatial resolution of STM), the $I–V$ measurement must be performed essentially at the same time as the topography measurement. This was first achieved by *Hamers* et al. [5.5], who utilized a multiplexing technique in which a voltage waveform was repetitively applied to the sample with a frequency of about 2 kHz (above the closed-loop bandwidth of the feedback control system). Of the 500 microseconds per waveform, $\simeq 400$ microseconds were devoted to the $I–V$ measurements, and during the remaining 100 microseconds the feedback controller operated normally in order to stabilize the sample–tip separation. By averaging a number of $I–V$ curves at each location, and by performing this measurement as the tip was slowly scanned across the surface, one generates a complete three-dimensional map of the surface. For each spatial location on the surface, there is a matching measurement of the tunneling $I–V$ characteristics, so that the electronic properties of the surface can be determined.

Analysis of tunneling spectroscopy is complicated by two facts. Firstly, the electronic density of states of the tip is usually unknown. Secondly, the voltage dependence of the tunneling probability is usually not known. The first issue is typically addressed by making comparisons between different locations on the same surface and by ensuring that all results are reproducible using different tips and different samples. Even though the electronic structure of the tip is unknown, it is at least *constant*, independent of spatial location. Thus, in tunneling spectra obtained at different locations on a surface with the same tip, the tip electronic structure contributes a constant background, but spatially dependent variations in the electronic structure can usually be determined free of tip effects. The second complication can also be addressed in several ways, depending on the details of the situation. In some cases, the effects of the voltage dependence of the tunneling probability can often be minimized by presenting the data as plots of $(dI/dV)/(I/V)$ vs. V (or equivalently, $d(\log I)/d(\log V)$ vs. V). In the WKB approximation, this is equivalent to:

$$\frac{dI/dV}{I/V} = \frac{\varrho_s(eV)\varrho_t(0) + \int_0^{eV} \frac{\varrho_s(E)\varrho_t(-eV + E)}{T(eV, eV)} \frac{dT(E, eV)}{dV} dE}{\frac{1}{eV} \int_0^{eV} \varrho_s(E)\varrho_t(-eV + E) \frac{T(E, eV)}{T(eV, eV)} dE}. \tag{5.3}$$

Feenstra et al. [5.4] have argued that since $T(eV, eV)$ and $T(E, eV)$ appear as ratios in the second term in the numerator and denominator, their dependences on separation and applied voltage tend to cancel. Thus, the normalization reduces the data to a form like

$$\frac{dI/dV}{I/V} = \frac{d(\log I)}{d(\log V)} = \frac{\varrho_s(eV)\varrho_t(0) + A(V)}{B(V)}, \tag{5.4}$$

which by definition is equal to unity at $V = 0$. Assuming that $A(V)$ and $B(V)$ vary slowly with voltage, this effectively normalizes that data. Unfortunately, on semiconductors this normalization is invalid because both $A(V)$ and $B(V)$ tend to vary rapidly with voltage, particularly at the band edges. In addition, the numerator in (5.3) vanishes at the band edges, making the normalized data very distorted near the band edges. Due to the shape of the tunneling barrier, for voltages greater than $\simeq 0.5$ V the tunneling electrons will typically arise from a relatively wide band $\simeq 0.3$ eV wide below E_F. As a result, the tunneling transmission probability cannot vary rapidly with voltage changes of less than a few tenths of a volt. For many studies aimed at understanding the density of states near band edges then, the voltage dependence of the tunneling barrier over this narrow voltage range can simply be ignored, and plots of I vs. V, dI/dV vs. V, or $d(\log I)$ vs. V are presented instead.

5.2 Scanning Tunneling Microscopy/Spectroscopy on Surfaces

5.2.1 Clean Group IV Semiconductors

a) Si(111)–(7 × 7) and Related Structures

Ever since the first atomic-resolution STM images were acquired by *Binnig*, *Rohrer* and coworkers [5, 6], the (7 × 7) reconstruction of Si(111) has been the *de facto* standard surface for STMs operating in ultrahigh vacuum. Although the structure of the (7 × 7) surface was not solved using STM alone, Binnig and Rohrer's initial observation of 12 adatoms provided an important clue which ultimately led to the widely accepted dimer adatom stacking fault (DAS) model proposed by *Takayanagi* et al. [5.7].

Clean Si(111)–(7 × 7) surfaces generally have a very low density of defects and present a very beautiful symmetry, consisting of two triangular subunits. Each triangular subunit contains six protrusions at the locations of the adatoms of the DAS model. At the corner holes where six triangular subunits join, the STM images show a depression about 2 Å lower than the height of the adatoms. The reconstruction is also accompanied by a number of surface states lying within the bulk bandgap. As a result, Fig. 5.1 shows that the STM topographic images of this surface are dependent on the sample bias voltage, since the voltage determines which electronic states can participate in tunneling [5.8].

When STM images are acquired with a negative sample bias so tunneling occurs from occupied electronic states (Fig. 5.1a), the faulted half of the unit cell appears higher than the unfaulted half. Additionally, within each half of the unit cell the three adatoms which are nearer the corner holes appear higher than the three interior adatoms. Thus, the images reveal four kinds of electronically inequivalent adatoms. At most positive sample bias voltages where tunneling occurs into unoccupied states of the sample (Fig. 5.1b), all twelve adatoms look nearly identical. The only exception to this is in a narrow voltage range near + 1.4 eV, in which the half of the unit cell without the stacking fault appears slightly higher than the faulted half.

The electronic energy spectrum of the (7 × 7) unit cell has been studied in more detail using local tunneling spectroscopy measurements [5.5, 9, 10]. Figure 5.2a shows the tunneling conductance measured at particular, well-defined locations within the (7 × 7) unit cell. At small negative bias voltage, nearly all the tunneling arises from the twelve adatoms, but is unevenly distributed. The conductance is higher in the faulted half than in the unfaulted half, and within each half it is higher on the three adatoms adjacent to the corner hole than on the three interior adatoms. When the voltage is increased beyond about −0.8 V, the conductance above the six rest atoms increases rapidly. From these measurements, it is possible to directly assign the geometric origin of the various surface states of Si(111)–(7 × 7). To compare the STM results in more detail with other experimental techniques, Fig. 5.2b shows the results of

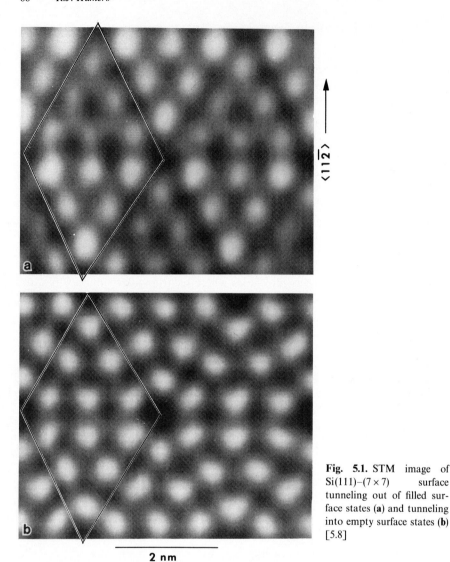

Fig. 5.1. STM image of Si(111)–(7 × 7) surface tunneling out of filled surface states (**a**) and tunneling into empty surface states (**b**) [5.8]

2 nm

photoemission and inverse photoemission spectroscopy by *Himpsel* and *Fauster* [5.11]; the significant point here is that wherever peaks are observed in the photoemission spectrum, sharp increases or "onsets" are observed in the tunneling conductance data, providing a 1 : 1 correspondence between the two types of measurements. Figure 5.2c shows normalized tunneling spectroscopy data averaged over many (7 × 7) unit cells. Clearly, there is uniform good agreement between the energy of the surface state observed in STM and those observed in photoemission spectroscopies.

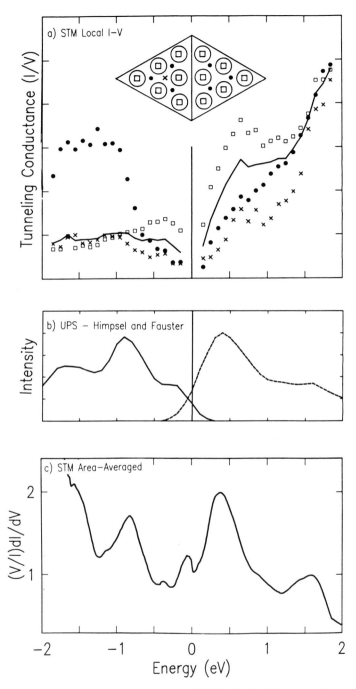

Fig. 5.2a–c. Tunneling spectroscopy on Si(111)–(7 × 7). (**a**) Local conductance measurements at specific points within unit cell. (**b**) Surface states observed in photoemission and inverse photoemission. (**c**) Normalized tunneling spectroscopy data averaged over entire unit cell [5.5]

Measurements made at specific locations within the unit cell have revealed the atomic origins of these states. Local tunneling spectroscopy measurements show that the adatoms of the (7×7) unit cell give rise to substantial tunneling currents even at very low bias, indicating that they have a high density of states near E_F and are the origin of the metallic surface state previously observed in photoemission studies [5.11]. The tunneling measurements show that this state has its highest state density around 0.15 eV below E_F, but the state density is higher in the faulted half than in the unfaulted half of the unit cell, and the state density is higher on the adatoms near the corner hole than on the interior adatoms. Because tunneling always occurs primarily from the occupied state nearest the Fermi level, topographic images acquired at negative bias voltage contain substantial contributions from this state and always show these asymmetries. This state also appears to be associated with the long-range order of the surface and is strongly affected by small amounts of contamination or disorder.

The surface state at -0.8 eV arises from the six rest atoms in each unit cell, while the state at -1.8 eV arises from the backbonds between the adatoms and the first full atomic layer. The unoccupied state at $+0.5$ eV is strongly localized on the twelve adatoms. The higher lying unoccupied state at $+1.5$ eV appears to be more uniformly distributed throughout the unit cell, but the energy of this state in the unfaulted half appears to be slightly lower than the faulted half; as a result, tunneling images acquired in a narrow range around $+1.5$ eV sample bias show an enhanced tunneling from the unfaulted half of the unit cell. This was also observed by *Becker* et al. [5.10] using a modulation technique.

On larger distance scales, Si(111)–(7×7) exhibits some interesting morphology. Steps on the Si(111)–(7×7) are usually very straight and tend to pass through the corner holes of the (7×7) unit cells, where the faulted and unfaulted halves join. The thermodynamic stability of the steps also has strong influences on the shapes of growth islands during molecular beam epitaxy as shown by *Köhler* et al. [5.12]. Figure 5.3 shows epitaxial islands of Si(111)–(7×7) grown at 725 K (top) and 825 K (bottom) [5.12]. The islands have a strong tendency to grow in a triangular shape, with the edges of the islands coinciding with the corner hole locations; unique reconstructions are observed on the small islands whose sizes are on the order of the size of the (7×7) unit cell.

In addition to the well-known (7×7) reconstruction, similar dimer-adatom-stacking fault structures have been observed in samples which were disordered as a result of laser annealing. *Becker* et al. [5.13] showed that the laser-annealed (1×1) surface is a result of a high degree of local disorder induced by the rapid quenching of the surface following the laser pulse. Tunneling spectroscopy measurements showed that the decreased long-range order was accompanied by the elimination of the adatom-derived surface state near -0.2 eV, but other states remained.

In laser-annealed and sputter-annealed silicon surfaces, *Becker* et al. [5.13, 14] also observed other local structures including small regions having (5×5), (9×9), and (2×2) symmetries. Recent STM studies of Si(111) surfaces grown by low-temperature molecular beam epitaxy [5.12] also show regions of (3×3),

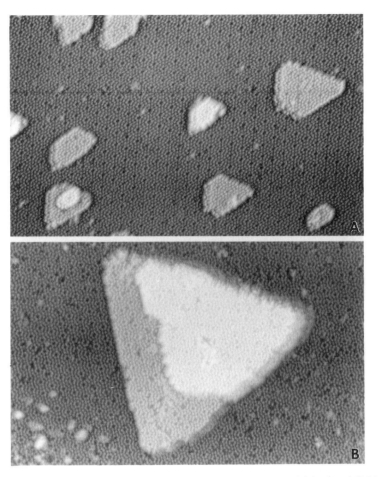

Fig. 5.3. STM image showing formation of triangular-shaped islands of Si(111)–(7 × 7) after epitaxial growth at 725 (*top*) and 825 (*bottom*) K [5.12]

(5 × 5), (9 × 9), and (2 × 2) symmetry, depending on the local Si coverage and the substrate temperature, while *Berghaus* et al. [5.15] observed a ($\sqrt{3} \times \sqrt{3}$) reconstruction on highly stepped Si(111) surfaces with terraces narrower than the width of a (7 × 7) unit cell. The DAS model for the (7 × 7) can be easily extended to include unit cells of (2n + 1 × 2m + 1) symmetry, where m and n are integers, so that the (3 × 3), (5 × 5), and (9 × 9) reconstructions are believed to be DAS-like, while the (2 × 2) and ($\sqrt{3} \times \sqrt{3}$) reconstructions are non-DAS. The (5 × 5) reconstruction appears to be only slightly less stable than (7 × 7). As shown in Fig. 5.4, low-temperature epitaxial growth of Si on Si(111)–(7 × 7) produces large regions with (5 × 5) symmetry coexisting with regions of (7 × 7) symmetry. STM images at negative sample bias show an asymmetry between the two

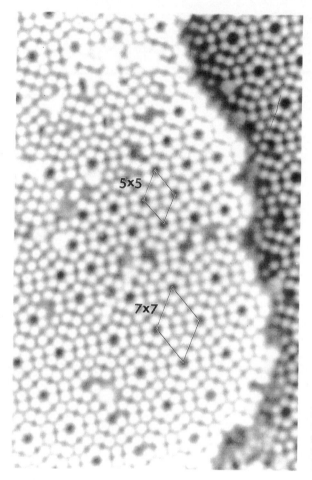

Fig. 5.4. STM images showing coexisting regions of (7×7) and (5×5) reconstructions on epitaxial islands [5.12]

triangular (5×5) subunits indicating the presence of a stacking fault under one half, as would be expected for a DAS-like reconstruction.

The (2×2) structure is essentially a subunit of the (7×7) unit cell containing one adatom and one rest atom. However, extended regions of (2×2) cannot have the alternating stacking-fault structures characteristic of the (5×5), (7×7), and (9×9) structures. The $\sqrt{3}$ structure [5.15] most likely arises from a Si adatom atop the three-fold sites of a bulk-terminated Si(111) lattice. These structures are observed primarily in very small islands of epitaxial silicon, in very disordered laser-annealed surfaces, or on very narrow step terraces where strain and size constraints prohibit the formation of alternating stacking faults.

b) Si(111)–(2 × 1)

The (2×1) reconstruction of Si(111) is readily produced by cleavage. STM images of this surface by *Feenstra* et al. [5.4] show chain-like structures; the images acquired at positive and negative bias show a shift of $\simeq 1.9\,\text{Å}$ (or half unit cell distance) along the $[01\bar{1}]$ direction and an additional shift of $0.7\,\text{Å}$ along the $[211]$ direction. These shifts directly correspond to the expected locations of the filled and empty sites predicted for the π-bonded chain reconstruction model proposed by *Pandey* [5.24]. Further support for Pandey's π-bonded chain model is provided by tunneling spectroscopy measurements (Fig. 5.5), which show peaks at -1.1, -0.3, 0.2, and 1.2 eV, in good correspondence with the peaks predicted from a one-dimensional tight-binding model for the π-bonded chains. Tunneling measurements of the decay length of the wavefunctions into the vacuum as a function of applied bias also have revealed interesting behavior resulting from dispersion of the surface-state bands, which disperse toward E_F as the parallel momentum $k_\parallel$ increases. At low voltages, tunneling can only occur from states with large $k_\parallel$, which have a short decay length. At higher voltages, tunneling occurs from states with large $k_\parallel$, which have the slowest decay. As shown in Fig. 5.5a, from measurements of the inverse decay length as a function of voltage, the dispersion of the surface-state bonds could be inferred from the tunneling measurements [5.4].

The (2×1) surface is thermodynamically unstable with respect to the (7×7) dimer-adatom-stacking fault reconstruction. Nevertheless, the (2×1) reconstruction has also been observed in several unusual circumstances. *Pashley* et al. [5.16] showed that (7×7) surfaces which were damaged by sputter annealing gave rise to small domains of (2×1) symmetry. Regions of (2×1) symmetry have also been observed in STM as a result of the rapid evaporation of submonolayer coverages of aluminum from Si(111), presumably as a kinetic intermediate.

STM images of steps on the Si(111)–(2×1) surface [5.17] indicate that many steps show evidence of a rebonding similar to the π-bonded reconstruction observed on flat terraces.

c) Si(001)

As illustrated in Fig. 5.6, the Si(001) surface reconstructs via a pairing of adjacent atoms of the (1×1) bulk-terminated lattice into dimers; this usually forms a (2×1) unit cell, although early LEED work also observed c(4 × 2) and p(2 × 2) patterns. *Appelbaum* et al. [5.18] showed that the electronic structure of such a dimer is described by a π state below E_F and an unoccupied π^* state above E_F. As shown in Fig. 5.7, bias-dependent STM images directly reflect the spatial distribution of the π-bonding state just below E_F and the π^*-antibonding state just above E_F. At negative sample bias, the (2×1) unit cell appears to consist of a single bean-shaped protrusion. At positive sample bias, however, two well-resolved protrusions are observed in each unit cell, separated by a deep depression. This depression arises because the wavefunction of the unoccupied π^*-antibonding state has a nodal plane along the center of the dimer bond, so that the density of unoccupied states vanishes there [5.18]. Figure 5.8 shows tunneling spectroscopy measurements [5.19, 20], revealing a surface-state band-gap of $\simeq 0.7\,\text{eV}$, with peaks at $-0.85\,\text{eV}$ and $+0.35\,\text{eV}$ corresponding to the

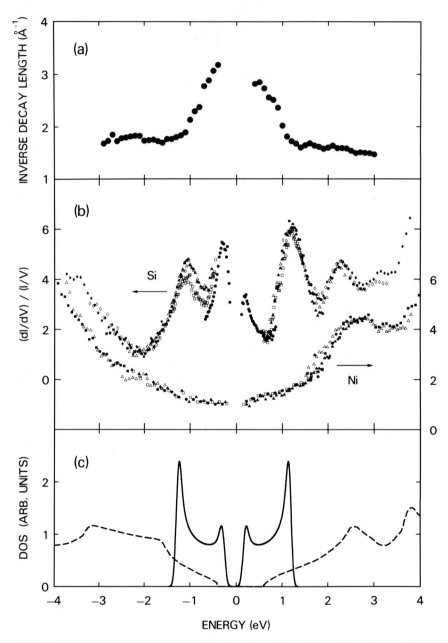

Fig. 5.5a–c. Spectroscopy and barrier height measurements on Si(111)–(2 × 1). (a) Variation in inverse decay length with applied voltage resulting from dispersion of surface-state bands. (b) Normalized tunneling spectra acquired at different sample–tip separations (overlayed), revealing the surface states of the (2 × 1) reconstruction. (c) Calculated density of states for Pandey's π-bonded chain [5.4]

Si(100)

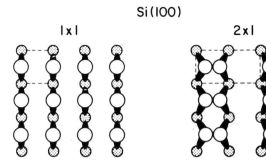

Fig. 5.6. Model for reconstruction of the bulk-terminated (1×1) lattice of the Si(001) surface into a (2×1) reconstruction via dimerization of adjacent atoms

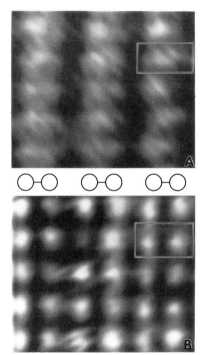

Fig. 5.7. STM images of Si(001) surface at negative sample bias, tunneling out of occupied surface states (*top*) and at positive bias, tunneling into unoccupied surface states (*bottom*) [5.19]

dimer π-bonding and π^*-antibonding states predicted by *Appelbaum* et al. [5.18].

As shown in Fig. 5.9, the surface always appears to have a high density of surface defects. These defects play important roles both because of the electronic energy spectrum and the effects they have on the nearby surface. The STM images show that in defect-free regions of the surface the dimers appear to lie in the surface plane, while near defects and steps a tilting of the dimers can often be observed. Such tilted dimers near defects readily lead to the higher order $c(4 \times 2)$

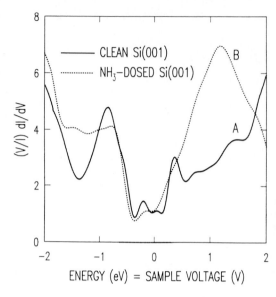

Fig. 5.8. Tunneling spectroscopy results on clean Si(001) surface and on Si(100)–(2 × 1)H hydrogen-terminated surface produced by NH_3 adsorption [5.19]

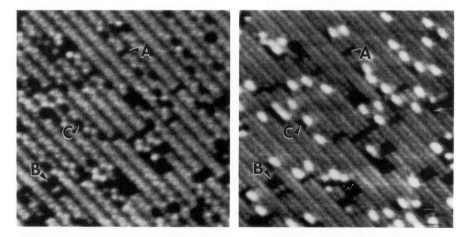

Fig. 5.9. STM images of identical regions of Si(001) surface at negative bias (*right*) and positive bias (*left*). A vacancy defect is labeled A and a Fermi level-pinning C defect is labeled C [5.23]

and p(2 × 2) reconstructions depending on whether the dimers in adjacent rows have the same phase or the opposite phase, respectively. Although band structure calculations [5.21] have suggested that the lowest energy (2 × 1) configuration may arise from tilted dimers, the predicted energy barrier between the two equivalent (2 × 1) configurations is sufficiently small that at room temperature, the dimers are most likely dynamically tilting between the two equivalent configurations, making them appear symmetric. Only near defects, steps (to be discussed below) and other local irregularities where a particular tilting directly is favored, can dimer buckling be observed at room temperature.

Since the (2 × 1) surface shows a surface-state bandgap, the surface Fermi energy position is strongly affected by local defects. Although photoemission studies have detected evidence for metallic behavior on Si(001) surfaces [5.22] *Hamers* and *Köhler* [5.23] used local tunneling spectroscopy measurements to show that such metallic behavior originated at particular kinds of defects. Using bias-dependent STM imaging and local tunneling spectroscopy measurements, the appearance of various types of atomic-sized defects, and their local tunneling I–V characteristics were probed. The two most common characteristic defects can be observed in Fig. 5.9. Individual dimer vacancies (labeled "A" in Fig. 5.9) show semiconducting behavior, in agreement with the π-bonded defect model of *Pandey* [5.24], and appear as vacancies both at negative and positive sample bias. The other characteristic C-shaped defect (labeled "C" in Fig. 5.9) appears significantly different at negative and positive bias. Tunneling I–V curves measured at this defect [5.23] show strongly metallic tunneling behavior and demonstrate that this particular type of defect is responsible for pinning the Fermi level at the Si(001) surface.

Other types of reconstructions on Si(001) can result from surface contamination. *Niehus* et al. [5.25] showed that small amounts of nickel contamination (and most likely other transition metals also) induced *ordering* of the vacancy defects, leading to larger structures having (2 × 8) and (2 × 10) symmetry, which were also detected in LEED.

Steps on Si(001) surfaces have been a subject of intense study. On samples which are aligned close to the [001] direction, only single-height steps are observed. At single-height steps, the (2 × 1) reconstruction is forced to rotate by 90° due to the tetrahedral coordination, resulting in the presence of two equivalent (2 × 1) domains on most Si(001) surfaces. This can be seen in Fig. 5.10, which shows a very large-area STM image of a Si(001) surface [5.26]. *Hamers* et al. [5.27] showed that three different types of steps are observed. For steps along [1$\bar{1}$0] direction (type "B" steps), two types of single steps are possible – one in which the dimers forming the lower step edge participate in dimer bonding, and one in which they do not; both types occur with approximately equal frequency. For steps along the [110] direction (type "A" steps), only one type of step is observed. In this configuration, stress at the step edge causes the dimers on the upper terrace to strongly buckle in the direction minimizing the strain [5.27].

Step energies and strain energies are important in the morphology of Si(001) surfaces. The energy of type "A" step edges (running parallel to the rows of dimers on the upper terrace, and forming a ⟨110⟩ step riser) is lower than that of type "B" step edges (running perpendicular to the rows of dimers on the upper terrace). This was predicted theoretically by *Chadi* [5.28] and leads to several experimentally observable features. Figure 5.10 shows that steps on Si(001) alternate between smooth and rough; the low-energy type "B" steps tend to be smooth because the energy to create a kink (and hence a small region of type

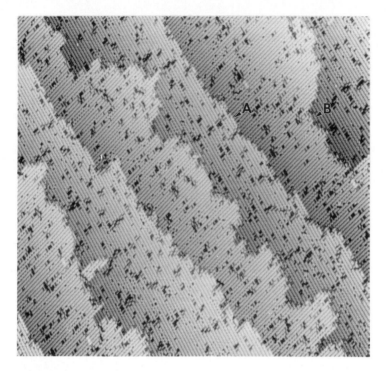

Fig. 5.10. Large-area STM image of Si(001). The alternating rough and smooth step edges are observed and are labeled A and B. This image shows a typical defect density of approximately 4% [5.26]

"A") is large; in contrast, type "A" steps tend to be rough, as the energy to create a small kink of type "B" is small. The (2×1) and (1×2) terraces which are separated by the single-height steps also have different strain energies, and *Swartzentruber* et al. [5.29] showed that straining the Si(001) sample can transform a two-domain surface into one which is predominantly a single domain [5.29].

On Si(001) samples tilted by several degrees toward $\langle 110 \rangle$, the surface tends to adopt a single-domain structure, with terraces separated by double-height steps. *Wierenga* et al. [5.30] first obtained STM images of these double steps, concluding that the observed pronounced dimer buckling was only consistent with a rebonding model proposed by *Chadi* [5.21]. Even on samples tilted by as much as 4°, however, studies both by *Griffith* et al. [5.31] and by *Swartzentruber* et al. [5.32] indicate that the surface never transforms completely to a single-domain structure. *Alerhand* et al. [5.33] calculated the equilibrium phase diagram for the (001) surface and showed that for small misorientations toward [110] single-height steps are energetically favored, while double-height steps are preferred for misorientations of larger than 2°. Samples tilted along the orthogonal [1$\bar{1}$0] direction do *not* show a single-domain structure due to the high

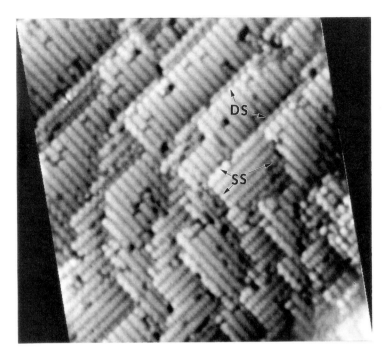

Fig. 5.11. STM image of off-cut Si(001) including both single-height steps (SS) and double-height steps (DS) [R.J. Hamers unpublished work]

energy associated with steps along this direction [5.28]. As shown in Fig. 5.11 [5.34], samples tilted at arbitrary angles between [110] and [1$\bar{1}$0] show a mixture of single- and double-height steps. Thus, although samples tilted precisely along [110] tend to show double steps, any slight tilting along [1$\bar{1}$0] causes the step edge to break into two single steps (one type A and one type B) [5.28], rather than forming the high-energy type A double step.

Studies of nucleation and growth processes on Si(001) surfaces show that both step energies and thermodynamics play important roles. Figure 5.12 shows STM images of Si(001) after molecular beam epitaxy of approximately 0.4 monolayer Si on Si(001). The resulting structures are extremely anisotropic, which appear to be a result of both thermodynamic and kinetic factors. At higher temperatures, *Hamers* et al. [5.26] showed that the anisotropy is reduced, but is still $\simeq$ 5:1, in the same direction anticipated on the basis of step energy calculations by *Chadi* [5.28]. *Mo* et al. [5.35, 36] showed that long annealing times reduced the anisotropy somewhat, indicating that the anisotropy at least in part arises from the growth kinetics. However, an aspect ratio of more than 3:1 was found to persist even after longer annealing times, indicating that step energies play at least a partial role in the observed growth anisotropy, but that extreme anisotropies most likely arise primarily from kinetic factors. The importance of kinetic factors is particularly apparent in the work of *Hoeven* et al.

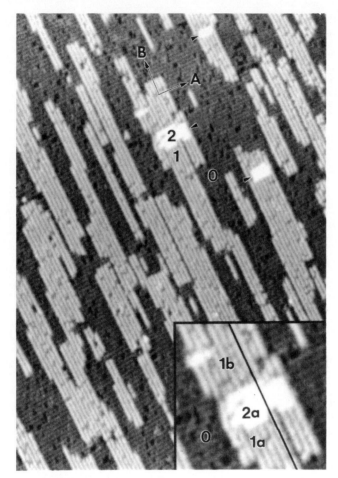

Fig. 5.12. Initial stages of epitaxial growth on Si(001), showing aniso-tropic growth. The inset shows how multilayer growth originates at an-ti-phase domain boun-daries [5.26]

[5.37, 38], who found that deposition of Si onto two-domain Si(001) surfaces at 750 K spontaneously produced surfaces which were predominantly one domain, due to preferential growth at the type "B" step edges (i.e., at the end of the dimer rows).

d) Si(110)

Experimental studies using diffraction techniques have identified a large number of ordered structures on this surface. Several of these have been observed in STM [5.39–42]. *Becker* et al. [5.42] observed a (5 × 1) reconstruction, while *Neddermeyer* and *Tosch* [5.41, 43] also observed structures with (2 × 5), (4 × 5), and (2 × 1) periodicities. The (2 × 5) unit cell appears to be the basic structural unit for the reconstructed surface, with an internal structure which appears as a chain-like structure running along the [$\bar{1}$10] direction. Orientation of these unit cells between adjacent chains naturally leads to the larger (4×5) unit cell observed also in LEED, while defects within the chain-like structures disturb the phase of the chains, and give rise to a local (2×1) order.

Recent experiments by *van Loenen* et al. [5.39] have indicated that the (4×5), (2×5), and (2×1) reconstructions may be the result of extremely small amounts of metal contamination. On scrupulously clean surfaces, they found that only a (16×2) reconstruction is observed; this unit cell is not rectangular, and so is more properly described in matrix notation as a $\begin{pmatrix} 1 & 17 \\ 2 & 2 \end{pmatrix}$ reconstruction. Figure 5.13 depicts the Si(110) surface and this unit cell. STM images of this surface, as shown in Fig. 5.14, show narrow alternating arrays of high and low terraces separated vertically by $\simeq 2.0\,\text{Å}$, with terrace edges aligned along the $\langle 112 \rangle$ direction for the $\begin{pmatrix} 1 & 17 \\ 2 & 2 \end{pmatrix}$ domain and along the $\langle 1\overline{1}2 \rangle$ direction for the equivalent $\begin{pmatrix} -1 & 17 \\ 2 & 2 \end{pmatrix}$ domain. STM images of the (16×2) structure show a period of $50\,\text{Å}$ perpendicular to the terrace edges and $13\,\text{Å}$ along the $\langle 1\overline{1}2 \rangle$ direction. The vertical height change of $2.0\,\text{Å}$ separating adjacent terraces appears equal to the Si interlayer spacing of $1.9\,\text{Å}$. At high resolution, STM images (Fig. 5.14, left) begin to show structure within the unit cells, and at the very highest resolution (Fig. 5.14, right) the internal structure is showing zig-zag chains of atoms which appear to be identical both on the upper and the lower terraces.

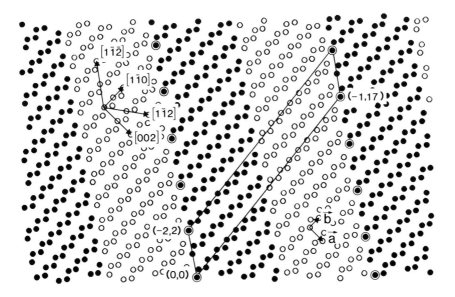

Fig. 5.13. Structural model of the Si(110) surface, with (16×2) unit cell outlined [5.40]

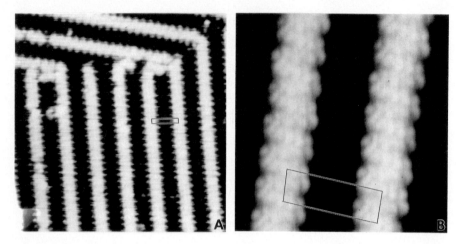

Fig. 5.14. High-resolution STM images of Si(110)–(16 × 2) surface; (**A**) image showing alternating array of terraces including domain boundary, (**B**) highest resolution, showing internal structure within zig-zag chains which form on each terrace [5.40]

When the terraces form an alternating up-down array, the (16 × 2) structure is formed. Alternatively, the terraces may step either up or down in a continuous fashion; this results in facets with (17 15 $\overline{1}$) orientation. Additionally, facets of the (15 17 3) family were observed.

Deposition of trace amounts of Ni and Cu onto the (16 × 2) structure produces the smaller unit cells described above. As little as 0.007 monolayer Ni destroys the order of the (16 × 2) reconstruction and leaves a (5 × 1)-type structure. Controlled deposition of 0.2 ML Ni, 0.06 ML Ni, and 0.01 ML Cu produced structures having (2 × 5), (7 × 5), and (5 × 1) periodicities, respectively.

e) Other Silicon Orientations

Berghaus et al. [5.9, 15] studied reconstructions of Si(112) and Si(223) surfaces. No well-ordered reconstructions with long-range order were observed. Instead, the surfaces exhibited large facets of (111) planes with the (7 × 7) and various other local reconstructions, which primarily occurred on rows and steps in the (110) direction. On the basis of their STM observations they concluded that the Si(112) surface has a low stability and tends to convert to larger (111)-oriented facets. On Si(223) they observed a five-fold periodicity along the ⟨110⟩ direction.

f) Ge(111)

STM studies of Ge(111) surfaces reveal some important differences but also some striking similarities to the Si(111) surface. *Becker* et al. [5.13, 42, 44] studied Ge(111) using STM and observed regions of c(2 × 8), (2 × 2), c(4 × 2), and

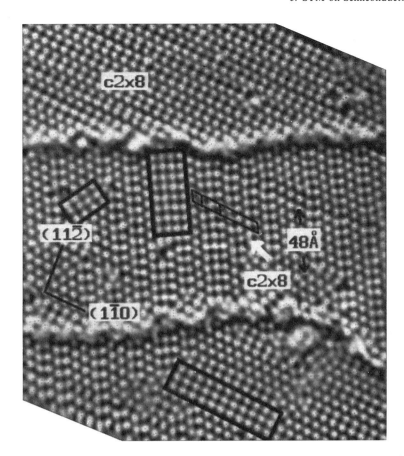

Fig. 5.15. STM image of Ge(111) surface, showing regions of (2×4) and $c(2 \times 8)$ symmetries [5.13]

$(\sqrt{3} \times \sqrt{3})$ symmetry. Figure 5.15 shows an STM image which includes regions of $c(2 \times 8)$ and (2×2) local symmetry. From these images, *Becker* et al. concluded that all four reconstructions could be most readily explained as ordered arrangements of Ge adatoms atop a bulk-like (111) lattice. They were also able to determine that the adatoms were located on those three-fold sites with a second layer germanium atom directly beneath, the T_4 geometry. Thus, the basic structural unit consists of a bulk-like Ge(111) lattice with one Ge adatom and one three-fold coordinated rest atom. The larger (2×2), $c(2 \times 8)$, and $c(4 \times 2)$ unit cells then arise from particular arrangements of this small unit.

The local (2×2) structure found on Ge(111) is identical to the local (2×2) unit cell formed by Si(111) as part of the larger (7×7) reconstruction – both contain a single adatom and a single rest atom in each unit cell. Unlike Si, however, the Ge(111) surface does not spontaneously form larger unit cells with stacking faults. Tunneling spectroscopy measurements on the Ge(111)–$c(2 \times 8)$ reconstruction show occupied states near -1.6 and -0.8 eV and unoccupied states near $+0.4$ and $+1.5$ eV. These energies are close to those previously observed on Si(111)–(7 × 7) [5.5], suggesting that the local atomic geometry is similar in both cases, with the -1.6 eV state arising from the Ge backbonds and the -0.8 eV state arising from the single three-fold coordinated rest atom in each (2 × 2) structural unit. In the case of Si, the 1.5 eV state is perturbed by the stacking fault so that its energy is different in faulted and unfaulted halves, while Ge shows only a single state near this energy.

g) Ge(001)

STM images of Ge(001) by *Kubby* et al. [5.45] show that the Ge(001) surface reconstructs by pairing of atoms along the $\langle 1\bar{1}0 \rangle$ direction to form dimers, in a manner similar to the Si(001) surface [5.27]. On both these surfaces, the axis of each dimer may lie parallel to the (001) surface plane (symmetric), or they may be tilted (buckled) out of the plane, as predicted by *Chadi* [5.21]. On Ge(001), many of the dimers appear to be symmetric, producing a (2 × 1) reconstruction. When the dimers are tilted, or buckled, the direction of tilting switches from dimer to dimer along a given row, producing zig-zag structures much like those previously observed on Si(001) [5.27]. The phase of buckling in adjacent rows can then be in-phase or out-of-phase, leading to p(2 × 2) and c(4 × 2) periodicities, respectively. Whereas on Si(001) dimer buckling is associated with defects

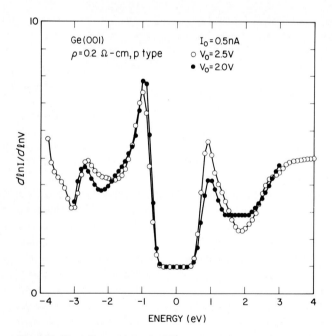

Fig. 5.16. Tunneling spectroscopy measurements on Ge(001), showing surface states at -2.7 eV, -0.9 eV, and $+0.9$ eV [5.45]

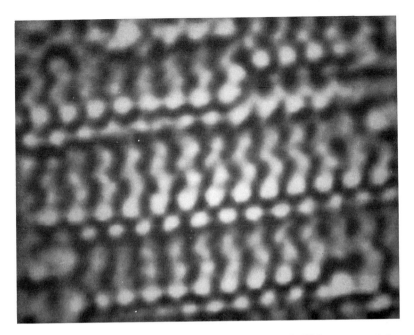

Fig. 5.17. STM image showing double-height steps on miscut Ge(001); pronounced dimer buckling is observed even on the flat terraces [5.46]

[5.27], on Ge(001) *Kubby* et al. [5.45] could find no apparent difference between the regions with (2×1) reconstructions and those with $c(4 \times 2)$ or $p(2 \times 1)$ reconstructions.

The electronic structure of this surface was investigated through tunneling spectroscopy measurements shown in Fig. 5.16 [5.45]. Two occupied surface states at 0.9 eV and 2.7 eV below E_F, and an unoccupied surface state at 0.9 eV above E_F are observed. Two sets of data obtained under slightly different tunneling conditions are shown.

On vicinal Ge(001) surfaces [5.46], double-height steps are observed when the sample is miscut by $\simeq 5.4°$ along the $[1\bar{1}0]$ direction, as shown in Fig. 5.17. Here the zig-zag buckled dimers can be observed, separated by double-height steps to form a single-domain (2×1) reconstruction.

h) Group IV Semiconductor Heteroepitaxy and Alloys

Becker and co-workers [5.42] have investigated a number of alloys of group IV semiconductors using both topographic imaging and tunneling spectroscopy. Comparing the tunneling spectra of Si(111)–(7×7) and SnGe(111)–(7×7), for example, Becker found that both surfaces have unoccupied surface states 1.4 eV above E_F in the unfaulted half and 1.6 eV and 2.8 eV above E_F in the faulted half. Thus, the electronic structure appears to be much more sensitive to the atomic geometry than to the identity of the constituent atomic species.

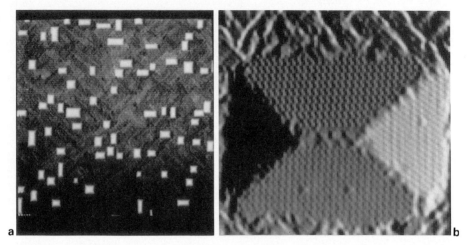

Fig. 5.18. STM images of germanium islands grown in Si(001). (**a**) 2000 × 2000 Å² region, showing growth of rectangular germanium islands after the first two monolayers of Ge epitaxy. (**b**) High-resolution image of individual Ge island. The base is rotated 45° with respect to the dimer rows of the substrate, and the sides of the rectangular pyramid are {105} facets [5.47]

Mo et al. [5.47] have studied the heteroepitaxy of germanium on Si(001) surfaces, which proceeds via a Stranski-Krastanov mechanism. As shown in Fig. 5.18, after the first two monolayers Ge forms small islands with rectangular bases rotated 45° from those of the substrate. The individual islands are commonly rectangular pyramids, with the sides comprised of {105} facets.

5.2.2 Clean Compound Semiconductor Surfaces

a) GaAs(110)

Of the compound semiconductors, the (110) face of GaAs has been most widely studied. Since Ga is electropositive and As electronegative, charge transfer between surface atoms leads to a high density of occupied states around the As atoms and a high density of unoccupied states around the Ga atoms. As shown in Fig. 5.19 [5.48], STM images obtained at positive sample bias then reveal the positions of the unoccupied states concentrated on the Ga atoms, while images at negative bias reveal the occupied states concentrated on the As atoms.

In general, apparent height differences between chemically inequivalent atoms cannot be directly interpreted in STM due to the influence of local electronic structure. By combining the STM results with linear augmented plane wave (LAPW) calculations, *Tersoff* et al. [5.49] showed that the lateral shift between the positive- and negative-bias maxima could be compared with that shift predicted from the LAPW calculations to determine an approximate buckling angle, as shown in Fig. 5.20. This comparison predicted that the

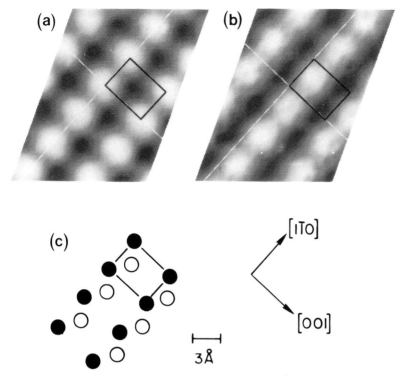

Fig. 5.19. STM image of GaAs(110) surface at positive (**a**) and negative (**b**) sample bias, together with ball model (**c**). Filled circles represent gallium atoms, open circles arsenic atoms [5.48]

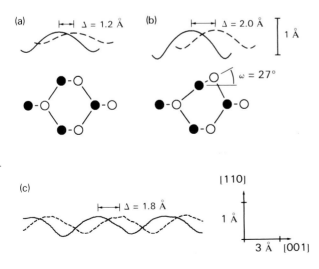

Fig. 5.20a–c. Calculated STM tip contours for occupied states (top, dashed line) and unoccupied (top, filled line) surface states of GaAs(110) at different buckling angles, compared with experimental contours (bottom); filled circles represent Ga atoms, open circles As atoms [5.49]

buckling angle was most likely between 29° and 31°, although the overall uncertainty would permit values between 23° and 35°.

Möller et al. [5.50] investigated the atomic structure of cleavage facets and found that steps occur preferentially along the [001], [112], and [114] directions. For [001] steps, the energetics associated with re-bonding the newly exposed [110] step edge play an important role, since [001] steps were observed to occur only in even multiples of the interlayer spacing. In contrast, steps along the [112] (exposing the As-terminated (111) face) and along the [114] directions occur in both even and odd multiples of the interlayer spacing.

b) GaAs(100)

In electron diffraction, this surface exhibits several different reconstructions. The gallium-rich face shows a (4×2) or $c(8 \times 2)$ reconstruction, and the arsenic-rich face shows a (2×4) or $c(2 \times 8)$ reconstruction. *Pashley* et al. [5.51] have studied the structure of the arsenic-rich face on GaAs(100) samples which were grown by molecular-beam epitaxy, capped with a thick layer of arsenic for protection, and finally cleaned under ultrahigh vacuum conditions by evaporating the

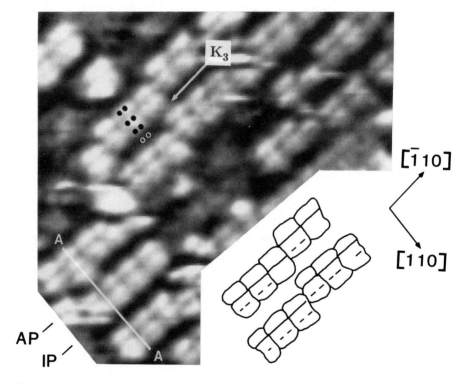

Fig. 5.21. STM image of empty states on GaAs(100) surface; the basic structural unit consisting of three As dimers and one vacancy is marked [5.51]

arsenic cap. The bulk-truncated GaAs(100) surface consists of a square array of arsenic atoms. As shown in Fig. 5.21, STM images reveal that these arsenic atoms pair together, forming surface dimers. While various models had proposed dimer formation and suggested that they might form the larger (3×4) and $c(2 \times 8)$ either by tilting and twisting of the dimers [5.52], Pashley's results show that the larger unit cells are constructed through vacancy formation as proposed by *Chadi* [5.53]. Pashley's STM measurements show that the arsenic dimers group together to form a structural sub-unit consisting of three As dimers and a single dimer vacancy, as outlined in Fig. 5.21. Ordering of these 4-dimer units then leads to extended regions of (2×4) or $c(2 \times 8)$ symmetry. Figure 5.22 schematically depicts various possible arrangements of this structural unit and the resulting longer range order which results.

More recent studies of *Biegelsen* et al. [5.54], using in situ molecular beam epitaxy demonstrate much improved surface order. In agreement with Pashley's earlier results they observe arsenic dimerization and the formation of 4-dimer units containing three As dimers and one dimer vacancy, with the (2×4) and $c(2 \times 8)$ unit cells formed by ordering of these structural units. In some cases, (2×4) unit cells consisting of two As dimers and two vacancies were observed; admixtures of these two possible local (2×4) unit cells allows for the (2×4) structure to exist over a wide range of surface stoichiometries. Biegelsen also observed a $c(4 \times 4)$ arrangement; like the (2×4) and $c(2 \times 8)$ reconstructions, the $c(4 \times 4)$ reconstruction arises from a four-dimer structural unit containing three dimers and a dimer vacancy; however, in the $c(4 \times 4)$ reconstruction the axis of each As–As dimer is rotated by $90°$ from its orientation in other reconstructions.

c) GaAs(111)

The gallium-terminated GaAs(111)-A surface exhibits a (2×2) reconstruction, studied by *Haberern* and *Pashley* [5.55]; Fig. 5.23 shows an STM image obtained while tunneling into the unoccupied states of this surface. Due to the charge transfer between Ga and As, the unoccupied surface states are expected to be localized on the Ga sites, in analogy with the GaAs(110) surface. STM images at positive sample bias are thus interpreted as revealing the locations of the Ga atoms. Superimposing a bulk-terminated lattice on the STM image shows that protrusions in the STM images corresponded to the locations of the bulk-terminated Ga sites, leading Haberern and Pashley to conclude that the (2×2) reconstruction arose from gallium vacancies in the outermost layer, as depicted in Fig. 5.24. The proposed structure is also in agreement with an electron-counting argument to achieve an uncharged surface.

d) GaAs($\overline{111}$)

Biegelsen et al. [5.56] have studied both the As-rich (2×2) and Ga-rich $(\sqrt{19} \times \sqrt{19})$ reconstructions of the arsenic-terminated GaAs($\overline{111}$) surface.

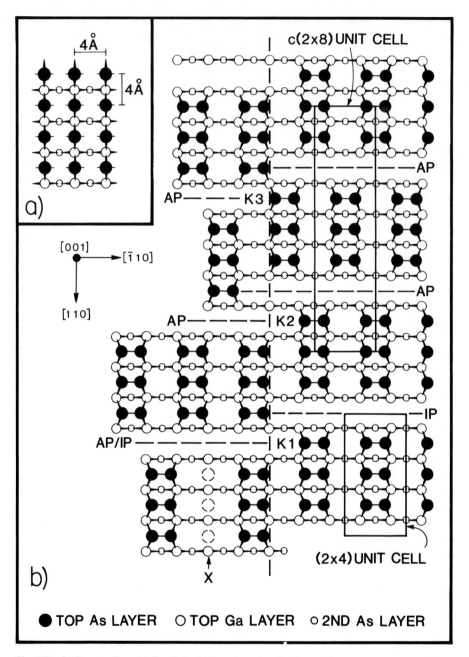

Fig. 5.22a, b. Structural model for GaAs(100) surface, showing how different arrangements of the 4-dimer structural unit result in (2×4) and $c(2 \times 8)$ symmetries [5.51]

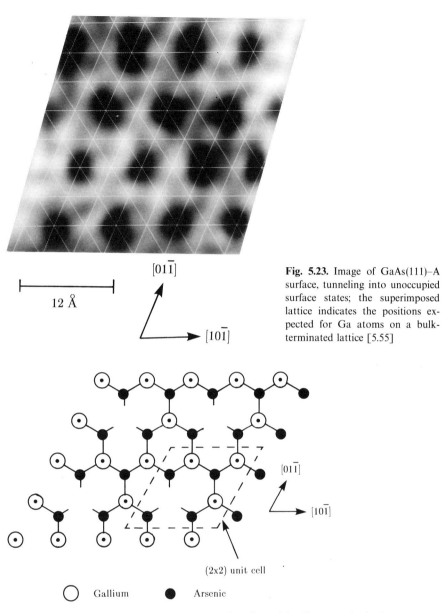

$[01\bar{1}]$

$[10\bar{1}]$

12 Å

Fig. 5.23. Image of GaAs(111)–A surface, tunneling into unoccupied surface states; the superimposed lattice indicates the positions expected for Ga atoms on a bulk-terminated lattice [5.55]

$[01\bar{1}]$

$[10\bar{1}]$

(2x2) unit cell

○ Gallium   ● Arsenic

Fig. 5.24. Proposed model for GaAs(111)–A surface, formed by Ga vacancies in the outermost surface layer [5.55]

STM images of the unoccupied and occupied states of the (2 × 2) reconstruction look identical, indicating that the protrusions accurately reflect the atomic positions of the surface atoms. Stacking fault boundaries were commonly observed, while high-resolution images showed one asymmetric protrusion per unit cell. On the basis of the STM images alone, Biegelsen and coworkers were unable to definitively establish whether the protrusions were adatoms or adatom clusters. Using detailed measurements combined with self-consistent pseudopotential calculations, they concluded that the protrusions arise from ordered arrays of arsenic trimers.

The $(\sqrt{19} \times \sqrt{19})$ reconstruction arises when the (2×2) reconstruction is heated. On the basis of the STM images, *Biegelsen* et al. [5.56] proposed a model in which the primary structural unit consists of six arsenic atoms in the top layer and twelve Ga atoms in the second layer, forming hexagonal rings.

e) GaAs–AlGaAs Heterostructure Interfaces

Salemink et al. [5.57] and *Albrektsen* et al. [5.58] used local tunneling spectroscopy measurements to study changes in the electronic structure across an AlGaAs–GaAs heterostructure interface. As shown in Fig. 5.25, STM topographic images show contrast between the GaAs and AlGaAs regions because of their different electronic structures. While the GaAs region appears very uniform, the AlGaAs layer has some apparent height modulations on the 15–20 Å length scale which may arise from variations in the local composition. The AlGaAs is also much more chemically reactive and shows some white protrusions which are most likely adsorbed oxygen and/or water vapor. The tunneling I–V curves measured on this structure were significantly different on the GaAs and AlGaAs regions; detailed analysis of the I–V curves obtained $\simeq 50$ Å away from the heterostructure interface enabled Albrektsen to measure the bandgaps of each region and also to determine the valence band offset. Nearer the interface, the electronic structure was more complicated. The heterostructure was designed such that the conduction band exhibited a depletion region (a local

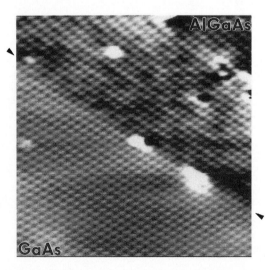

Fig. 5.25. STM image of GaAs–AlGaAs heterostructure. The heterostructure was grown along the (001) plane and then cleaved to expose the (110) plane [5.58]

increase in the conduction band energy) on the AlGaAs side of the interface and a confinement region on the GaAs side. As a result, tunneling into the empty states at the interface showed a strong enhancement in current at the confinement layer and a strong decrease at the depletion region, extending $\simeq 10$ nm away from the interface itself.

5.2.3 Adsorbates and Overlayers on Semiconductors

a) Group III Metals on Silicon

Group III transition metals on Si(111) have been extensively studied by several groups [5.59, 60]. At 1/3 monolayer, Al, Ga and In all form ordered $(\sqrt{3} \times \sqrt{3})$ structures which arise from simple adatom structures, in which the metal atop adsorbs on top of a bulk-like Si(111) lattice in the so-called T_4 site. Additionally, these $(\sqrt{3} \times \sqrt{3})$ overlayers all show the same characteristic defect, which *Hamers* [5.59] showed as arising from having a Si adatom (instead of a group III metal adatom) in the T_4 site. Hamers performed detailed tunneling spectroscopy studies both of the "ideal" $(\sqrt{3} \times \sqrt{3})$ overlayer and directly at these defects, and compared their electronic characteristics with theoretical calculations. Interestingly, the electronic structures of Al and Si adatoms are sufficiently different that they can be separately imaged; the Al adatom has a pronounced unoccupied surface state about 1.1 eV above E_F, while the Si atom has an occupied surface state just below E_F; as a consequence, imaging at positive sample bias reveals the Al adatoms, while imaging at negative bias reveals the Si adatoms.

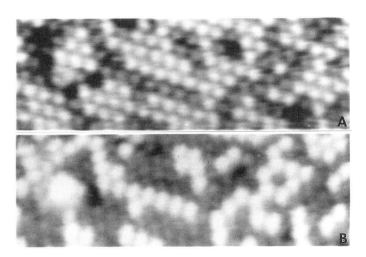

Fig. 5.26. STM image of 1/6 monolayer Al on Si(111), forming a $\sqrt{3}$ structure with 50% Al and 50% Si adatoms. The top image shows the empty states, revealing the Al adatoms; the bottom image shows the filled states, revealing the Si adatoms [5.61]

This can be clearly observed in Fig. 5.26, which shows an STM image of Si(111) with only 1/6 monolayer Al deposited, so that half the adatoms are expected to be Al, and half Si.

While Al, Ga and In all behave similarly at low coverage, at higher coverage they diverge, forming a variety of reconstructions. Most of these reconstructions exhibit strongly voltage-dependent changes indicative of electronic states which are both spatially and energetically separated. An example is shown in Fig. 5.27, which shows images of the $(\sqrt{7} \times \sqrt{7})$ overlayer of Al on Si(111), which occurs at 3/7 monolayer Al. Using voltage-dependent STM imaging, tunneling spectroscopy, and lattice superposition analysis techniques, *Hamers* [5.61] showed that this structure arises from clusters of Al adatoms on two-fold sites. The Al atoms are imaged at positive sample bias (top), while negative bias images reveal a single Si dangling bond in each unit cell (bottom). *Nelson* and *Batra* [5.62]

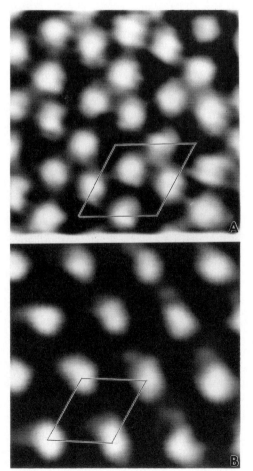

Fig. 5.27. STM image of $(\sqrt{7} \times \sqrt{7})$ overlayer of Al on Si(111), demonstrating strong voltage-dependent symmetry changes resulting from strongly localized electronic states. Top image shows empty states, revealing Al adatoms on two-fold sites; bottom image shows filled states, revealing position of silicon dangling bond [5.59]

calculated the electronic density of states for this geometry and found it to be very low in energy and with empty and filled surface states concentrated in the same locations observed experimentally. Other Al-induced structures having $(\sqrt{3} \times \sqrt{3})$, $(\sqrt{7} \times \sqrt{7})$, (7×7), and (2×1) symmetries were studied by *Hamers* [5.59, 61, 63, 64] using STM and tunneling spectroscopy. From STM voltage-dependent imaging, the adsorption sites were identified for these superstructures; by combining this with tunneling spectroscopy, it was possible to directly correlate the local bonding geometry with the local electronic structure, which was supported by a variety of electronic structure calculations.

At higher coverages, In shows a number of reconstructions including $(\sqrt{31} \times \sqrt{31})$, $(\sqrt{7} \times \sqrt{3})$, and (4×1) symmetries which were observed by *Nogami* et al. [5.65, 66]. At coverages approaching one monolayer, Ga/Si(111) exhibits an interesting incommensurate structure which is approximately (6.3×6.3). In addition to images presented by *Park* et al. [5.60], a detailed study was performed by *Chen* et al. [5.67], who find that the incommensurate nature of this overlayer is essentially a stress effect. They propose that gallium and silicon form two nearly coplanar interpenetrating lattices, forming an inert graphite-like structure. The overall periodicity of this structure (6.3 lattice constants, or about 24 Å) apparently arises from misfit dislocations due to the different lattice constants of this Ga-Si overlayer structure and the bulk Si lattice. Similar misfit dislocations were also found for Al overlayers on Si(111), which near 1 monolayer forms a relatively disordered structure with an average period of near 7 lattice constants [5.61]. Both Al and Ga overlayers show apparently ragged boundaries defining comparatively featureless triangular-shaped regions with an *average* spacing of 6.3 (for Ga) or $\simeq 7.0$ (for Al) lattice constants.

Like the other group III elements, boron on silicon forms a $(\sqrt{3} \times \sqrt{3})$ structure at 1/3 monolayer coverage; however, the structure of this $\sqrt{3}$ structure is very different from the simple T_4 adatom structure formed by the other members of this series. Early studies by *Dumas* et al. [5.68, 69] indicated that the symmetry along the short diagonal of the $(\sqrt{3} \times \sqrt{3})$ unit cell was broken. Based on these observations, they proposed a model in which the boron atom *substitutes* into the first normal full silicon layer, leaving a silicon adatom on top of the surface. Later studies by *Bedrossian* et al. [5.70] and by *Lyo* et al. [5.71] did not observe this asymmetry. These later studies indicate that the *B* atom does indeed substitute into the outermost layer of Si atoms, but the substitutional site is the atom immediately beneath the T_4 adatom, rather than the site originally proposed by *Dumas*.

On the Si(001) surface, *Baski* et al. [5.72] and *Nogami* et al. [5.73] found that Ga adsorption resulted in the formation of Ga–Ga dimers, and that these Ga dimers arrange themselves to form regions of (2×3) symmetry below 0.3 ML and (2×2) symmetry at 0.5 ML coverage. At very low coverage, the Ga dimers formed very long rows perpendicular to the dimer rows of the Si surface, with aspect ratios of more than 20:1.

b) Group V Metals on Silicon

The adsorption of the group V elements As and Sb on silicon and germanium has received much attention due to the valence electronic structure and possible application to heteroepitaxy of III–V semiconductors on group IV semi-conductors. *Becker* et al. [5.74] and *Copel* et al. [5.75] studied arsenic adsorption on Si(111) and observed locally ordered (1 × 1) regions 50–100 Å in diameter, but found that the surface had a large amount of disorder on longer length scales. *Becker* et al. performed a detailed analysis of both the geometric and electronic structure of this surface, showing that the STM images were consistent with arsenic substitution for the Si in the outermost half-double layer of the (111) lattice. Tunneling spectra showed a surface-state gap of 1.9–2.3 eV, with the edges of the surface state bands at 0.6 eV above and 1.1 eV below E_F, in agreement with theoretical quasiparticle calculations [5.76]. Both *Copel* [5.75] and *Becker* et al. [5.74, 76] observed the highest corrugation with a sample bias near + 0.6 V, corresponding to the edge of the empty surface-state band. Theoretical calculations [5.76] showed that the empty surface state had its highest density near the Si atoms, leading *Becker* to conclude that the STM images reflected the positions of the silicon atoms, not the arsenic atoms.

On the Si(001) surface, *Becker* et al. [5.76] found that arsenic adsorption preserved the (2 × 1) symmetry but produced a surface with much greater perfection than the starting Si(001) surface. Using miscut wafers which produce a nearly single-domain (2 × 1) reconstruction, *Becker* showed that at low adsorption temperatures the arsenic adsorbs on top of the Si(001) dimers, forming a new terrace with dimers rotated by 90° from the original substrate. At higher temperatures, the arsenic dimers were able to displace the outermost Si layer, producing a (1 × 2) reconstruction with domains rotated by 90° with respect to those formed at lower temperatures.

Rich et al. [5.77] studied Sb adsorption on Si(001) using both STM and core-level photoemission spectroscopy. Their STM results show that the Sb atoms adsorb on the Si(001) dimers, saturating the dangling bond. This contributes additional electron density to the previously empty π^* antibonding state, pushing it below the Fermi energy. As a result, the occupied dimer states show a clear node in the center; these changes are similar to those observed after H-termination of the dimer dangling bonds [5.19]. At positive sample bias, the Sb-covered surface shows only a single protrusion at the position of the dimer center.

c) Transition Metals on Silicon

Palladium on Si(111) was studied by *Köhler* et al. [5.78] and exhibits a surprising spatial selectivity. STM images revealed that after depositing palladium on Si(111) at room temperature, 95% of the palladium atoms were located on the *faulted* half of the (7 × 7) unit cell. This remarkable spatially selective adsorption is present even at higher coverages, up to nearly one monolayer where the Pd islands begin to cover the entire (7 × 7) unit cell.

Room temperature deposition of silver on Si(111) also occurs preferentially in the faulted half, as observed in STM measurements by *Tosch* et al. [5.79]. At higher coverages, silver forms triangular flat islands which appeared to be a single plane of close-packed silver atoms. At higher temperatures, silver on Si(111) forms an ordered $(\sqrt{3} \times \sqrt{3})$ structure which has been extensively studied; unlike the simple $\sqrt{3}$ unit cell produced by group III metals, the Ag-induced $\sqrt{3}$ unit cell has a honeycomb shape, with two protrusions in each unit cell. Two simultaneous STM studies [5.80, 81] of this system made similar experimental observations, but came to opposite conclusions. Based on electron counting and the observation of a semiconducting electronic structure, *van Loenen* et al. [5.80] concluded that the Ag atoms formed an embedded honeycomb and that the surface atoms were Si adatoms, while *Wilson* and *Chiang* [5.81] concluded that the observed protrusions were the Ag atoms themselves. A later study [5.82] showed that the protrusions were located at sites corresponding to three-fold sites of the Si(111) lattice, but the atomic geometry of the $\sqrt{3}$ Ag/Si(111) surface remains in question.

Ni adsorption on Si(111) forms a $(\sqrt{19} \times \sqrt{19})R23.4°$ structure studied by *Wilson* and *Chiang* [5.81], who found that this reconstruction arises from a single subsurface Ni atom in each unit cell, with this Ni atom bonded to six silicon adatoms. This surface tends to be relatively disordered, most likely due to the fact that the Ni atom is not just a simple adatom structure and activation barriers for diffusion are then very high. An incommensurate overlayer is also produced by the adsorption of Cu on Si(111); *Demuth* et al. [5.83] utilized two-dimensional Fourier analysis technique to relate the short- and long-range periodicities observed in STM to the diffraction spots observed in LEED.

Silver on Si(001) has also been reported to show asymmetric growth. At very low coverage, *Hashizume* et al. [5.84] found that the Ag atoms also formed rows of dimers oriented perpendicular to the underlying dimer rows. At higher coverage, *Brodde* et al. [5.85] found that the Ag atoms formed structures aligned *parallel* to the dimer rows of the underlying substrate, although on this surface there were a large number of vacancy defects aligned into a (2×8) reconstruction, similar to that observed by *Niehus* et al. [5.25] on a Ni-contaminated surface. Both Hashizume and Brodde reported instabilities in the tunneling current near Ag atoms and observed diffusion of Ag atoms during the measurements.

d) Alkali Metals on Silicon

Hashizume et al. [5.86] have studied the adsorption of K and Li on Si(001). Surprisingly, the alkalis adsorb on top of the dimers, aligning themselves in rows perpendicular to the dimer rows of the underlying substrate. While this is the same orientation as has been observed for adsorption of Ag and Ga on Si(001), it is contrary to the predictions which had been made previously for the alkalis.

e) Metals Adsorbed on Ge Surfaces

In comparison with silicon surfaces, little work has been done studying adsorption on Ge surfaces. *Becker* et al. [5.74, 76] studied the adsorption of arsenic on Ge(111) and Ge(100) surfaces. As on the analogous Si(111) surface, arsenic adsorption on the Ge(111) surface produces a (1×1) structure. The (1×1) domains were approximately 100–150 Å in extent, separated by trenches which were three lattice constants wide. The trenches apparently act as a strain relief mechanism arising from the 2% mismatch between the As–Ge and Ge–Ge bond lengths. Tunneling spectroscopy measurements revealed surface states 0.7 eV below and 0.2 eV above E_F, in agreement with theoretical quasiparticle calculations [5.87].

f) Adsorption on Compound Semiconductor Surfaces

The GaAs(110) surface has been by far the most widely studied compound semiconductor surface. *Stroscio* et al. [5.88] studied oxygen adsorption on GaAs(110) and showed that charge transfer between the O and GaAs surfaces produced band-bending effects; these band-bending effects are visible as a smooth change in the apparent surface extending out radially from the adsorbed oxygen atom, with an effective range determined by the bulk electrostatic screening length.

Virtually all the remaining work studying adsorbates on GaAs has focussed on the electronic properties and the mechanism of Schottky-barrier formation. *Martensson* and *Feenstra* [5.89] studied the adsorption of antimony on GaAs(110). The geometry of the surface antimony atoms is similar to the geometry of the silicon atoms in the π-bonded chain model of the Si(111)–(2×1) surface, except that the dangling bonds are half-filled on the Si surface but are completely filled on the Sb/GaAs surface. There are two inequivalent Sb atoms per unit cell; they observed both atoms at low bias voltages, but at high voltage only the Sb atoms bonded to the underlying Ga atoms were observed. *Ludeke* et al. [5.90] studied the adsorption of bismuth on GaAs(110). At coverages between 0.5 and 1 monolayer, they found that the Bi atoms formed zig-zag chains which were located both between and on top of the GaAs chains. Near 1 monolayer coverage the bismuth atoms formed dislocations spaced $\simeq 25$ Å apart, which gave rise to unoccupied acceptor states within the bandgap which pin the Fermi level. *Feenstra* [5.91] studied Au and GaAs(110) and found that the behavior of Au was similar to that of Sb and Bi; all three generate empty states which act as acceptors and filled states which act as donors.

First et al. [5.92] studied cesium on GaAs(110) and found a remarkable ordering of the Cs atoms, forming long one-dimensional chains. High-resolution images of these chains show that they are composed of a repeating unit of two inequivalent Cs atoms; this inequivalence arises from the underlying GaAs lattice and is not a spontaneous symmetry-breaking. At higher coverage, chains consisting of three inequivalent Cs atoms are formed. The formation of such close-packed ordered structures indicates that charge transfer between Cs and GaAs is smaller than originally predicted, since large amounts of charge transfer would produce significant repulsive interactions between the Cs adatoms. *First* et al. [5.93] also performed detailed tunneling spectroscopy studies of iron clusters on GaAs(110). Iron clusters containing approximately 13 atoms were shown to be non-metallic, while clusters with more than 35 atoms showed indications of metallic character and a well-defined Fermi energy. In the GaAs surrounding these clusters, a continuum of Fe-induced states were observed. Measurements of the (lateral) spatial extent of these states showed that the length correlated with the energy of the state, with states at mid-gap having the smallest effective range and states near the conduction- and valence-band edges having the longest decay length. This is a consequence of wavevector matching of the complex (decaying) wavevectors within the gap with the continuum of states existing on the large Fe clusters.

5.2.4 Chemical Reactions on Semiconductor Surfaces

Several chemical reactions between gas-phase molecules and the Si(001) surface have been studied with STM. *Hamers* et al. [5.19, 94] utilized STM and tunneling spectroscopy to study the dissociation of NH_3 on Si(001). The dissociation leaves hydrogen on the surface, which saturates the dangling bonds of the Si(001) dimers. Whereas on the clean surface tunneling at negative sample bias occurred primarily at the center of the Si–Si dimer bond, dimers which had reacted with the hydrogen showed tunneling primarily from the *ends* of the dimer bonds. This was interpreted in terms of the different spatial location of the occupied π-state of the clean surface and the occupied Si–H related states on the reacted dimers. Tunneling spectroscopy was also used to study the changes in overall electronic state density associated with the reaction. *Avouris* and *Wolkow* [5.95] later studied the reaction of NH_3 with the Si(111)–(7 × 7) surface. Here, it was found that the rest atoms have higher chemical reactivity than the adatoms, and the reactivity of the four kinds of inequivalent adatoms of the (7 × 7) reconstruction correlate with their local density of states at the Fermi energy [5.5, 8].

Oxidation of the Si(111) surface has also been a popular subject. *Leibsle* et al. [5.96] studied the initial stages of oxidation and found that missing adatom defects acted as nucleation sites for the oxidation. *Lyo* et al. [5.97] used STM, tunneling spectroscopy and electronic structure calculations to study the oxidation using both molecular oxygen and also N_2O as oxidizers. In the very early stages of oxidation, they observed two kinds of reacted sites at positive sample bias; in one, the adatoms appeared to be higher than normal (unreacted) (7 × 7) adatoms, and in the other they appeared to be lower. Electronic structure calculations for various geometries predicted that only the structure consisting of a single oxygen atom inserting into a Si–Si backbond produced a high density of states above E_F, leading them to assign this structure to the high sites. This kind of site was also preferentially located at sites corresponding to adatoms next to the corner holes. They proposed that this was a minority species, however, and that the predominant oxidation mechanism starts with a precursor state in which an O_2 molecule attaches to the adatom of a single silicon adatom, followed by an arrangement to insert one of the oxygen atoms into a backbond while leaving one attached to the dangling bond. *Pelz* and *Koch* [5.98] also studied the oxidation using both STM and tunneling spectroscopy; they also observed both low and high reacted sites, with tunneling spectra nearly identical to those of *Lyo* et al. [5.97]. However, Pelz and Koch were also able to directly image the same area as a function of oxygen exposure directly observe a two-stage mechanism which disagrees with the mechanism proposed by *Lyo* et al. [5.97]. Pelz and Koch showed that adatoms which appeared high at low oxygen exposure turned low upon further exposure. Their results indicated that the bright adatoms and dark adatoms are two different stages in a two-stage mechanism; oxygen atoms first insert into one backbond to form the high adatoms, and upon further exposure a second oxygen atom inserts into another backbond of the same Si adatom, forming the low adatom structure.

Villarubia and *Boland* [5.99] studied the reaction of atomic chlorine with Si(111)–(7 × 7). At small positive bias voltages, they found that the reacted sites of the Si(111)–(7 × 7) appeared dark, due to the elimination of the unoccupied adatom-related surface state lying 0.5 eV above E_F [5.5]; at large positive voltages, tunneling occurred into a broad, Cl-derived σ^* state. STM images taken after saturation Cl exposure and a 950 K anneal showed that the adatom layer of the (7 × 7) reconstruction was stripped away, leaving the underlying 42 atoms of the rest-atom layer exposed in each unit cell. They demonstrated that the underlying stacking fault was preserved, giving rise to a difference between the average height of each half and variations in the detailed distribution of tunneling current within each half of the unit cell.

5.3 Other Tunneling Techniques Applied to Semiconductors

5.3.1 Surface Photovoltage

Conventional tunneling spectroscopy provides information on the energies of the surface states relative to the Fermi level. However, tunneling spectroscopy does not provide any information about the electronic structure in the sub-surface space-charge region, extending 25–1000 Å below the surface; additionally, STM only provides information about the *equilibrium* electronic properties of the surface. By combining scanning tunneling microscopy with optical excitation, these limitations can be surpassed, and it is possible to glean new information about the electronic structure of the sub-surface region and to learn about carrier dynamics also.

The basic idea of surface photovoltage measurements is shown in Fig. 5.28. Due to presence of surface states lying within the bulk band gap, there is a

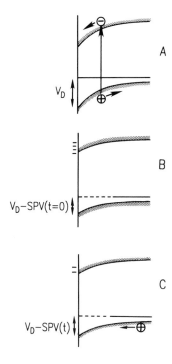

Fig. 5.28. Schematic representation of surface photovoltage effect on p-type semiconductor. (**A**) Optical illumination creates electron–hole pairs, which separate in subsurface space charge region. (**B**) Carrier motion produces an excess of surface charge (negative, at Si) which produces the photovoltage and reduces the band bending. (**C**) Recombination is limited by activation of holes over Schottky barrier, which occurs by thermionic emission of holes over a Schottky barrier of height V_D–SPV, where V_D is the barrier in the absence of illumination.

charge exchange between the surface and the bulk in order to maintain Fermi-Dirac population statistics in the surface layer and to maintain charge neutrality both at the surface and in the bulk. The net result of this charge exchange is that there is a strong electric field in the sub-surface region (as much as 10^6 V/cm for highly doped material), causing the valence and conduction band energies to bend in the near-surface region. On silicon surfaces, the high density of surface states in the gap tends to always pin the surface Fermi level in the middle of the gap, so that the band bending is always upward for bulk n-doped material and downward for bulk p-doped material. Illumination of the surface with above-bandgap light creates electron–hole pairs throughout its absorption depth. Those charge carriers created in the sub-surface space charge region experience a strong electric field, which tends to separate the electron and hole. If the bands bend downward (as is generally true for p-type material), then the electrons move toward the surface and the holes move away. The situation is reversed if the bands bend up, as on n-type material. The spatial separation of electrons and holes generates a voltage difference between the surface and the bulk, known as the surface photovoltage or SPV. We see from this discussion that measuring the sign of the photovoltage immediately reveals whether the sub-surface band bending is upward or downward, thus providing information on the sub-surface electronic structure.

Figure 5.29 shows surface topography (a) and spatially resolved surface photovoltage data (b) of *Hamers* and *Markert* [5.100] obtained using tunneling

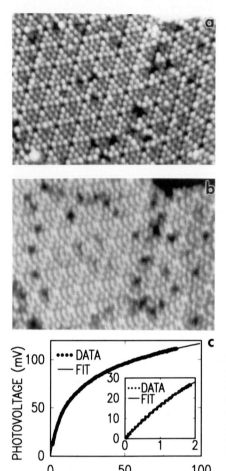

Fig. 5.29. Spatially resolved surface photovoltage measurements on Si(111)–(7 × 7). **(a)** shows surface topography, including (7 × 7) domain boundary; **(b)** shows spatially resolved surface photovoltage, demonstrating strong decreases in SPV near defects; **(c)** shows dependence of SPV on illumination intensity and a fit to a function of the form SPV − A log(1 + BI) [5.100]

potentiometry, demonstrating spatial variations in the photovoltage signal on the 20 Å-diameter length scale which are correlated with defects in the surface topography. On Si(001) surfaces, *Cahill* and *Hamers* [5.101] observed changes in the SPV near defects (most likely SiC pinning centers), with a characteristic length of $\simeq 120$ Å on 0.1 Ωcm material and 600 Å on 2.0 Ωcm material. Both on Si(111)–(7 × 7) and Si(001), the distance scale of the SPV variations is only about one third that expected from the bulk electrostatic screening length.

When the SPV is small compared to the magnitude of the band bending, then Hamers and Markert showed that the SPV was accurately fit by a function of the form: SPV = A log(1 + BI), where A is a constant, I is the illumination intensity, and B is a constant inversely proportional to the local recombination rate. The excellent agreement with the function is shown in Fig. 5.29. A similar functional form is predicted by a recent SPV theory of *Hecht* [5.102]; this theory treats the rate-limiting step in the recombination process as the thermionic emission of bulk majority carriers over the Schottky barrier created by the surface band bending. The theory that the prefactor A should be $\simeq kT/q = 26.1$ mV at room temperature, which agrees quantitatively with the value $\simeq 31$ mV measured experimentally [5.103] using tunneling potentiometry on Si(111)–(7 × 7). On Si(001) the prefactor is slightly higher, which can also be accounted for in an ad hoc manner in Hecht's theory through the introduction of a non-ideality factor which corrects for effects such as carrier recombination within the space-charge region. The SPV is very sensitive to the magnitude of the local band bending because the thermionic emission which controls the recombination process varies exponentially with the Schottky barrier height.

The spatial variations in SPV can arise from several factors. Surface inhomogeneities which produce a change in the local surface charge will pin the Fermi energy at a different energy within the bulk bandgap, and so the sub-surface band bending will be different from surrounding regions, producing a different photovoltage. If this mechanism is responsible, then one expects that the contrast should reverse between n-type and p-type material, since a defect which locally increases the band bending for one doping type should decrease it for the other. This behavior was observed by *Cahill* and *Hamers* [5.101] on Si(001) near defects believed to be silicon carbide clusters. In this case, the effect of the defect is to change the overall band bending, leading to a change in the local SPV.

A second contrast mechanism can arise because defects and surface in-homogeneities also modify the carrier transport kinetics. For example, a defect which introduces new states which mix with the conduction-band states (on n-type material) or with the valence-band states (on p-type material) will effectively lower the Schottky barrier which limits the surface recombination. In this scheme, the new electronic states associated with the defect modify the *kinetics* of the carrier transport in the near-surface region. This mechanism was proposed by *Hamers* and *Markert* [5.100] to explain their SPV results on Si(111)–(7 × 7), where intrinsic defects at boundaries between Si(111)–(7 × 7) domains strongly reduced the local SPV both on n-type and on p-type material.

In principle, it is possible to quantitatively determine both the magnitude of the band bending and the local recombination rate from such measurements. *Hamers* and *Cahill* [5.104] have recently investigated the time scale of the SPV effect using capacitance microscopy combined with picosecond optical excitation; as a result of the non-linear dependence of the SPV on the illumination intensity, optical correlation techniques can provide information on the carrier dynamics on fast time scales even using conventional slow STM electronics.

5.3.2 Tunneling-Induced Luminescence

Tunneling-induced luminescence measurements are in many ways complementary to the surface photovoltage measurements just described. Instead of using photons to inject carriers and then using the STM to probe these carriers (as in SPV), tunneling-induced luminescence measurements turn the situation around and use the STM tip to inject electrons or holes, and then measure the resulting recombination luminescence. While on indirect-gap semiconductors such as silicon the radiative recombination is very inefficient, on direct-gap semiconductors such as GaAs radiative recombination is extremely efficient, providing high photon count rates. *Abraham* et al. [5.105] utilized this technique to study (100)-oriented GaAs–AlGaAs heterostructures, cleaved in ultrahigh vacuum to expose the (110) surfaces. The light emitted into a solid angle of $\simeq 0.2$ steradian was collected, filtered, and imaged onto a GaAs-cathode photomultiplier tube. The optical system was optimized for photon energies of 1.56 ± 0.33 eV. In the GaAs regions, tunneling into the conduction band produced high fluxes of emitted photons, several thousand photons collected per second for each nanoamp of tunneling current. In contrast, tunneling into the AlGaAs layers typically produced low photon fluxes, with the contrast between GaAs and AlGaAs decreasing as the sample voltage was increased between 2 V and 5 V. Quantum wells as thin as 10 nm were imaged in this manner, with the luminescence efficiency abruptly changing over a distance of less than 2 nm. The energy of the emitted photons was peaked at the direct gap between the GaAs conduction band minimum and the valence band maximum, irrespective of whether the photons were injected into the GaAs or AlGaAs layers, thus indicating that at higher voltages the luminescence observed when tunneling into the AlGaAs layers arises from photons diffusing into the GaAs region before radiatively combining. At the lowest sample voltages (where the contrast is highest) electrons injected into the AlGaAs layer are trapped in a thin confinement layer in the AlGaAs near the GaAs–AlGaAs interface and do not diffuse into the GaAs region. This technique holds great promise for studies of carrier transport in various types of semiconductor heterostructures.

5.3.3 Potentiometry

In order to study charge transport processes in semiconductor heterostructures with the STM, *Muralt* and *Pohl* [5.106] implemented the first tunneling *potentiometry* apparatus. This is a three-terminal measurement, in which a current is passed through the sample of interest while the STM tip is used to measure the potential as a function of position. At semiconductor heterostructures, large voltage drops can occur over very short distances – on the order of 1 V over 100 Å for a typical p–n junction. Utilizing this potentiometry apparatus, it is possible to directly study the potential changes occurring in the junction region. Muralt and Pohl invented this technique and applied it to semiconductor hetero-structures. At GaAs–AlGaAs interfaces cleaved under ultrahigh vacuum conditions, topographic images showed only small changes in the surface height at the interface; however, simultaneous potentiometry measurements showed a very sharp change in the potential [5.107]. On a more complicated structure consisting of 1000 Å layer of GaAs sandwiched between n-type AlGaAs and p-type AlGaAs layers, *Muralt* et al. [5.108] were able to show how the thickness of the recombination layer changes as a function of the bias. The three-terminal nature of scanning tunneling potentiometry makes it a powerful technique for studying electrical transport in small-scale semiconductor structures.

5.3.4 Ballistic Electron Emission Microscopy (BEEM)

An important variation of STM which can be applied to the study of buried interfaces is ballistic electron emission microscopy, or BEEM [5.110]. The most important distinction between STM and BEEM is that while STM is a two-terminal measurement involving a sample and a tip, BEEM involves three electrodes: a bulk semiconducting sample, a thin metallic overlayer and the tip. Although several variations of the BEEM technique have been reported [5.110–5.112], the basic technique involves the tunneling of electrons from a tip into a metal overlayer on a semiconducting substrate. Electrons tunneling through the vacuum gap pass from the tip into the metal film, where most of them scatter and are collected at the second terminal. However, a small percentage of the electrons (1–5%) propagate through the metal film and pass into the semiconducting substrate, where they are collected at the third terminal. There is usually a potential (Schottky) barrier at the metal–semiconductor interface that the electrons must overcome in order to pass into the semiconductor. While the tunneling current from the tip to the metal film is maintained at a constant value, the bias between the metal film and the semiconducting substrate is varied, and the small current arising at the semiconductor from unscattered electrons is measured. The resulting I–V curve is characteristic of the electronic structure of the interface between the metal and the semiconductor.

Perhaps the most surprising aspect of BEEM is that even when tunneling through metal films of several hundred ångstroms thickness, spatial resolution on the order of 10 Å can be achieved. This high spatial resolution is a result of strict momentum-conservation laws which must be obeyed by the electrons in order to pass from the metal to the semiconducting region. These laws result in a critical angle only a few degrees away from the interface normal; electrons which impinge on the metal–semiconductor interface at angles larger than this critical angle are reflected back into the metal, analogous to the phenomenon of total internal reflection in optics.

The initial BEEM studies [5.110] focused on mapping spatial variations in the Schottky barrier height at buried Au/Si and Au/GaAs interfaces on nanometer distance scales. These measurements show that Au/Si interfaces exhibited only small spatial variations in the Schottky barrier height, Au/GaAs(100) surfaces showed pronounced variations on the order of 0.1 V. In a more complete analysis of the experimental BEEM spectra, *Bell* and *Kaiser* [5.111] developed a detailed theoretical model for the shape of the I–V curves which demonstrated excellent agreement with the experimental measurements. In the case of Au/GaAs interfaces, detailed analysis of the BEEM spectra reveals not only the energy of the conduction band minimum, but also the energies of other critical points in the band structure. Bulk GaAs has a relatively complicated conduction band; in addition to the conduction band minimum at Γ, satellite minima occur at the L and X points in the Brillouin zone. In the BEEM studies, plots of dI/dV vs. V show steps at 0.89, 1.18, and 1.36 eV, which are associated with ballistic electrons propagating through the interface into states at Γ, L, and X, respectively. Furthermore, at these energies the slope of the dI/dV vs. V curves can be related to the electron effective mass.

More recently, *Bell* et al. [5.112] have demonstrated that a modification of the BEEM technique can also be used to study ballistic *hole* transport across interfaces. This modified BEEM technique was demonstrated at Au/GaAs(100) and Au/Si(001) interfaces. While electrons can be easily injected from the tip into the metal overlayer, injecting holes is somewhat more involved. Several methods of hole injection were studied, involving different sample–tip polarity and semiconductor doping. When the tip is negatively biased, electrons first tunnel from the tip into the metal overlayer; through inelastic scattering processes, these hot electrons transfer energy to equilibrium electrons, exciting the equilibrium electrons to conduction band states above E_F and leaving a hole in the metal conduction band below E_F. These holes can propagate to the metal–semiconductor interface, and those holes propagating through the interface constitute a current flow through the semiconductor. When the tip is positively biased, electrons tunneling from the metal overlayer into the STM tip leave holes behind, which may directly undergo ballistic transport through the metal–semiconductor interface. By choosing the appropriate combination of semiconductor doping and tip polarity, it is thereby possible to study both the direct ballistic transport of holes and the production of holes by carrier-scattering processes. These studies provide new insight into transport, scattering processes, and hot-carrier creation.

References

5.1   J. Tersoff, D.R. Hamann: Phys. Rev. Lett. **50,** 1998 (1983)
5.2   J. Tersoff, D.R. Hamann: Phys. Rev. B **31,** 805 (1985)
5.3   R.J. Hamers: Ann. Rev. Phys. Chem. **40,** 531 (1989)
5.4   R.M. Feenstra, J.A. Stroscio, A.P. Fein: Surf. Sci. **181,** 295 (1987)
5.5   R.J. Hamers, R.M. Tromp, J.E. Demuth: Phys. Rev. Lett. **56,** 1972 (1986)
5.6   G. Binnig, H. Rohrer, C. Gerber, E. Weibel: Phys. Rev. Lett. **50,** 120 (1983)
5.7   K. Takayanagi, Y. Tanishiro, M. Takahashi, S. Takahashi: J. Vac. Sci. Tech. A **3,** 1502 (1985)
5.8   R.M. Tromp, R.J. Hamers, J.E. Demuth: Phys. Rev. B **34,** 1388 (1986)
5.9   T. Berghaus, A. Brodde, H. Neddermeyer, S. Tosch: J. Vac. Sci. Tech. A **6,** 483 (1988)
5.10   R.S. Becker, J.A. Golovchenko, D.R. Hamann, B.S. Swartzentruber: Phys. Rev. Lett. **55,** 2032 (1985)
5.11   F.J. Himpsel, T. Fauster: J. Vac. Sci. Tech. A **2,** 815 (1984)
5.12   U. Köhler, J.E. Demuth, R.J. Hamers: J. Vac. Sci. Tech. A **7,** 2860 (1989)
5.13   R.S. Becker, B.S. Swartzentruber, J.S. Vickers, T. Klitsner: Phys. Rev. B **39,** 1633 (1989)

5.14 R.S. Becker, T. Klitsner, J.S. Vickers: Phys. Rev. B **38**, 3537 (1988)
5.15 T. Berghaus, A. Brodde, H. Neddermeyer, S. Tosch: Surf. Sci. **181**, 340 (1987)
5.16 M.D. Pashley, K.W. Haberern, W. Friday: J. Vac. Sci. Tech. A **6**, 488 (1988)
5.17 R.M. Feenstra, J.A. Stroscio: Phys. Rev. Lett. **59**, 2173 (1987)
5.18 J.A. Appelbaum, G.A. Baraff, D.R. Hamann: Phys. Rev. Lett. **35**, 11 (1975)
5.19 R.J. Hamers, P. Avouris, F. Bozso: Phys. Rev. Lett. **59**, 2071 (1987)
5.20 R.J. Hamers, P. Avouris, F. Bozso: J. Vac. Sci. Tech. **6**, 508 (1988)
5.21 D.J. Chadi: Phys. Rev. Lett. **43**, 43 (1979)
5.22 P. Martensson, A. Cricenti, G. Hansson: Phys. Rev. B **33**, 8855 (1986)
5.23 R.J. Hamers, U.K. Köhler: J. Vac. Sci. Tech. A **7**, 2854 (1989)
5.24 K. Pandey: Phys. Rev. Lett. **47**, 1913 (1981)
5.25 H. Niehus, U.K. Köhler, M. Copel, J.E. Demuth: J. Microsc. **152**, 735 (1988)
5.26 R.J. Hamers, U.K. Köhler, J.E. Demuth: Ultramicroscopy **31**, 10 (1989)
5.27 R.J. Hamers, R.M. Tromp, J.E. Demuth: Phys. Rev. B **34**, 5343 (1987)
5.28 D.J. Chadi: Phys. Rev. Lett. **57**, 1691 (1987)
5.29 B.S. Swartzentruber, M.B. Mo, M.B. Webb, M.G. Lagally: J. Vac. Sci. Tech. A **8**, 210 (1990)
5.30 P.E. Wierenga, J.A. Kubby, J.E. Griffith Phys. Rev. Lett. **59**, 2169 (1987)
5.31 J.E. Griffith, G.P. Kochanski, J.A. Kubby, P.E. Wierenga: J. Vac. Sci. Tech. A **7**, 1914 (1989)
5.32 B.S. Swartzentruber, M.B. Mo, M.B. Webb, M.G. Lagally: J. Vac. Sci. Tech. A **7**, 2901 (1989)
5.33 O. Alerhand, A.N. Berker, J. Joannopoulos, D. Vanderbilt, R. Hamers, J. Demuth: Phys. Rev. Lett. **64**, 2406 (1990)
5.34 R.J. Hamers: unpublished work
5.35 Y.W. Mo, B.S. Swartzentruber, R. Kariotis, M.B. Webb, M.G. Lagally: Phys. Rev. Lett. **63**, 2393 (1989)
5.36 Y.W. Mo, R. Kariotis, B.S. Swartzentruber, M.B. Webb, M.G. Lagally: J. Vac. Sci. Tech. A **8**, 201 (1990)
5.37 A.J. Hoeven, D. Dijkkamp, E.J. van Loenen, J.M. Lenssinck, J. Dieleman: J. Vac. Sci. Tech. A **8**, 207 (1990)
5.38 A.J. Hoeven, J.M. Lenssinck, D. Dijkkamp, E J. van Loenen, J. Dieleman: Phys. Rev. Lett. **63**, 1830 (1989)
5.39 E.J. van Loenen, D. Dijkkamp, A.J.. Hoeven: J. Microscopy **152**, 487 (1989)
5.40 A.J. Hoeven, D. Dijkkamp: Surf. Sci. **211/212**, 165 (1989)
5.41 H. Neddermeyer, S. Tosch: J. Microscopy **152**, 149 (1988)
5.42 R.S. Becker, B.S. Swartzentruber, J.S. Vickers: J. Vac. Sci. Tech. A **6**, 472 (1988)
5.43 H. Neddermeyer, S. Tosch: Phys. Rev. B **38**, 5784 (1988)
5.44 R.S. Becker: Proc. Nat. Acad. Sci. USA **84**, 4667 (1987)
5.45 J.A. Kubby, J.E. Griffith, R.S. Becker, J.S. Vickers: Phys. Rev. B **36**, 6079 (1987)
5.46 J.E. Griffith, J.A. Kubby, P.E. Wierenga, R.S. Becker, J.S. Vickers: J. Vac. Sci. Tech. A **6**, 493 (1988)
5.47 Y.-W. Mo, D.E. Savage, B.S. Swartzentruber, M.G. Lagally: Phys. Rev. Lett. **65**, 1020 (1990)
5.48 R.M. Feenstra, J.A. Stroscio: J. Vac. Sci. Tech. B **5**, 923 (1987)
5.49 J. Tersoff, R.M. Feenstra, J.A. Stroscio, A.P. Fein: J. Vac. Sci. Tech. A **6**, 497 (1988)
5.50 R. Möller, R. Coenen, B. Koslowski, M. Rauscher: Surf. Sci. **217**, 289 (1989)
5.51 M.D. Pashley, K.W. Haberern, W. Friday, J.M. Woodall, P.D. Kirchner: Phys. Rev. Lett. **60**, 2176 (1988)
5.52 P.K. Larsen, J.F. Veen, A. Mazur, J. Pollmann, J.H. Neave, B.A. Joyce: Phys. Rev. B **26**, 3222 (1982)
5.53 D.J. Chadi: J. Vac. Sci. Tech. A **5**, 834 (1987)
5.54 D.K. Biegelsen, R.D. Bringans, J.E. Northrup, L. Swartz: Phys. Rev. B **41**, 5701 (1990)
5.55 K.W. Haberern, M.D. Pashley: Phys. Rev. B **41**, 3226 (1990)
5.56 D.K. Biegelsen, L.L. Swartz, R.D. Bringans: J. Vac. Sci. Tech. A **8**, 280 (1990)
5.57 H.W. Salemink, H.P. Meier, E. Ellialtiogly, J.W. Gerritsen, P.R. Muralt: Appl. Phys. Lett. **54**, 1112 (1989)

5.58   O. Albrektsen, D.J. Arent, H.P. Meier, H.W. Salemink: Appl. Phys. Lett. **57**, 31 (1990)
5.59   R.J. Hamers: J. Vac. Sci. Tech. B **6**, 1462 (1988)
5.60   S. Park, J. Nogami, C.F. Quate: J. Microscopy **152**, 727 (1988)
5.61   R.J. Hamers: Phys. Rev. B **40**, 1657 (1989)
5.62   J.S. Nelson, I.P. Batra: Proceedings of the NATO Advanced Research Workshop on Metallization and Metal–Semiconductor Interfaces (Plenum, New York, in press)
5.63   R.J. Hamers, J.E. Demuth: J. Vac. Sci. Tech. A **6**, 512 (1988)
5.64   R.J. Hamers, J.E. Demuth: Phys. Rev. Lett. **60**, 2527 (1988)
5.65   J. Nogami, S. Park, C.F. Quate: Phys. Rev. B **36**, 6221 (1987)
5.66   J. Nogami and S.I. Park: J. Vac. Sci. Technol. B **6**, 1479 (1988)
5.67   D.M. Chen, J.A. Golovchenko, P. Bedrossian, K. Mortensen: Phys. Rev. Lett. **61**, 2867 (1988)
5.68   P. Dumas, F. Phibaudau, F. Salvan: J. Microscopy **152**, 751 (1988)
5.69   F. Thibaudau, P. Dumas, P. Mathiez, A. Humbert, D. Satti, F. Salvan: Surf. Sci. **211–212**, 148 (1989)
5.70   P. Bedrossian, R.D. Meade, K. Mortensen, D.M. Chen, J.A. Golovchenko, D. Vanderbilt: Phys. Rev. Lett. **63**, 1257–60 (1989)
5.71   I.I. Lyo, K. Kaxiris, P. Avouris: Phys. Rev. Lett. **63**, 1261 (1989)
5.72   A.A. Baski, J. Nogami, C.F. Quate: J. Vac. Sci. Tech. A **8**, 245 (1990)
5.73   J. Nogami, S. Park, C.F. Quate: Appl. Phys. Lett. **53**, 2086 (1988)
5.74   R.S. Becker, R. Klitsner, J.S. Vickers: J. Microscopy **152**, 157 (1988)
5.75   M. Copel, R.M. Tromp, U.K. Köhler: Phys. Rev. B **37**, 10756 (1988)
5.76   R.S. Becker, B.S. Swartzentruber, J.S. Vickers, M.S. Hybertson, S.G. Louie: Phys. Rev. Lett. **60**, 116 (1988)
5.77   D.H. Rich, F.M. Leibsle, A. Samsavar, E.S. Hirschorn, T. Miller, T. Chiang: Phys. Rev. B **39**, 12758 (1989)
5.78   U.K. Köhler, J.E. Demuth, R.J. Hamers: Phys. Rev. Lett. **60**, 2499 (1988)
5.79   S. Tosch, H. Neddermeyer: Phys. Rev. Lett. **61**, 249–52 (1988)
5.80   E.J. van Loenen, J.E. Demuth, R.M. Tromp, R.J. Hamers: Phys. Rev. Lett. **58**, 373 (1987)
5.81   R.J. Wilson, S. Chiang: Phys. Rev. Lett. **58**, 369–72 (1987)
5.82   R.J. Wilson, S. Chiang: Phys. Rev. Lett. **59**, 2329 (1987)
5.83   J.E. Demuth, U.K. Köhler, R.J. Hamers, P. Kaplan: Phys. Rev. Lett. **62**, 641 (1989)
5.84   T. Hashizume, R.J. Hamers, J.E. Demuth, K. Markert, T. Sakurai: J. Vac. Sci. Tech. A **8**, 249 (1990)
5.85   A. Brodde, D. Badt, S. Tosch, H. Neddermeyer: J. Vac. Sci. Tech. A **8**, 251 (1990)
5.86   T. Hashizume, Y. Hasegawa, I. Kamiya, T. Ide, I. Sumita, S. Hyodo, T. Sakurai: J. Vac. Sci. Tech. A **8**, 233 (1990)
5.87   M.S. Hybertsen, S.G. Louie: Phys. Rev. Lett. **58**, 1551 (1987)
5.88   J.A. Stroscio, R.M. Feenstra, A.P. Fein: Phys. Rev. Lett. **58**, 1668 (1987)
5.89   P. Martensson, R.M. Feenstra: J. Microscopy **152**, 761 (1988)
5.90   R. Ludeke, A. Taleb-Ibrahimi, R.M. Feenstra, A.B. McLean: J. Vac. Sci. Tech. B **7**, 936 (1989)
5.91   R.M. Feenstra: Phys. Rev. Lett. **63**, 1412 (1989)
5.92   P.N. First J.A. Stroscio, R.A. Dragoset, D.T. Pierce, R.J. Celotta: Phys. Rev. Lett. **63**, 1416 (1989)
5.93   P.N. First, J.A. Stroscio, R.A. Dragoset, D.T. Pierce, R.J. Celotta: Phys. Rev. Lett. **63**, 1416 (1989)
5.94   R.J. Hamers, P. Avouris, F. Bozso: J. Vac. Sci. Tech. B **5**, 1387 (1987)
5.95   P. Avouris, R. Wolkow: Phys. Rev. B **39**, 5091 (1989)
5.96   F.M. Leibsle, A. Samsavar, T. Chiang: Phys. Rev. B **38**, 5780 (1988)
5.97   I. Lyo, P. Avouris, B. Schubert, R. Hoffman: J. Phys. Chem. **94**, 4400 (1990)
5.98   J. Pelz and R.H. Koch: Phys. Rev. B **42**, 3761 (1990)
5.99   J.S. Villarubia, J.J. Boland: Phys. Rev. Lett. **63**, 306 (1989)
5.100 R.H. Hamers, K.W. Markert: Phys. Rev. Lett. **64**, 1051 (1990)
5.101 D.G. Cahill, R.J. Hamers: J. Vac. Sci. Tech. B**9**, 564 (1991)

5.102 M.H. Hecht: Phys. Rev. B **41,** 7918 (1990)

5.103 R.J. Hamers, K.W. Markert: J. Vac. Sci. Tech. A

5.104 R.J. Hamers, D.G. Cahill: Appl. Phys. Lett. **57,** 2031 (1990)

5.105 D.L. Abraham, A. Veider, C. Schonenberger, H.P. Meier, D.J. Arent, S.F. Alvarado: Appl. Phys. Lett. **56,** 1564 (1990)

5.106 P. Muralt, D. Pohl: Appl. Phys. Lett. **48,** 514 (1986)

5.107 P. Muralt: Surf. Sci. **181,** 324 (1987)

5.108 P. Muralt, H. Meier, D.W. Pohl, H.W. Salemink: Appl. Phys. Lett. **50,** 1352 (1987)

5.109 Y. Kuk, P.J. Silverman: Appl. Phys. Lett. **48,** 1597 (1986)

5.110 W.J. Kaiser, L.D. Bell: Phys. Rev. Lett. **60,** 1406 (1988)

5.111 L.D. Bell. W.J. Kaiser: Phys. Rev. Lett. **61,** 2368 (1988)

5.112 L.D. Bell, M.H. Hecht, W.J. Kaiser: Phys. Rev. Lett. **64,** 2679 (1988)

6. STM on Layered Materials

R. Wiesendanger and *D. Anselmetti*

With 44 Figures

Layered materials [6.1] exhibit many properties which make them special with regards to STM studies and in scanning probe microscopy in general. Thus an entire chapter is devoted to STM investigations of layered materials, although, depending on the electrical conductivity of the layered materials, these investigations could also have been included in the chapters on metals or semiconductors. In contrast, insulating layered materials cannot be studied by STM but only by its relative, atomic force microscopy (AFM). Such investigations are reported in the chapter entitled "Scanning Force Microscopy" in the second volume on "Scanning Tunneling Microscopy".

Layered materials are characterized by the anisotropy of their physical properties which stems from their structural anisotropy. True layer-type or quasi-two-dimensional structures consist of neutral sandwiches held together only by weak van der Waals forces. Therefore, layered materials can easily be cleaved, providing atomically flat terraces of up to several thousand square ångstroms which is advantageous for atomic resolution STM studies. Furthermore, in these true layer-type compounds, cleaving does not create dangling bonds and the freshly cleaved surface may stay clean for a long time. Therefore, atomic resolution STM studies of layered materials have proven to be feasible even in air and liquids. Finally, the elastic properties of layered materials differ significantly from those of three-dimensional structures, layered materials being generally much softer. This property first led to the recognition of the influence of forces in STM experiments.

Materials with a layered structure may be classified into three categories of different structural complexity:

1. Layered materials which are built up by identical sheets. These sheets can be shifted relative to each other by a fraction of the in-plane lattice constant. For graphite, two types of layered structures exist: hexagonal graphite with an ABABAB ... stacking sequence and rhombohedral graphite with an ABCABC ... stacking sequence. Section 6.1 will be devoted to STM studies of the graphite surface.

2. Layered materials which are built up by a repetition of a sandwich of sheets. The individual sheets belonging to the sandwich are chemically different but remain simple structures. The transition metal dichalcogenides TX_2 (T = transition metal, X = chalcogenide) belong to this category. In this case, the sandwich consists of a top chalcogenide layer, a middle transition metal

layer and a bottom chalcogenide layer. Topographic STM studies of transition metal dichalcogenides will be described in Sect. 6.3. Some of these transition metal dichalcogenides exhibit a collective electronic transport phenomenon, the charge density waves (CDW). STM has already contributed much to our present knowledge of the local aspects of the CDW state. This will be discussed in Sect. 6.4. For completeness, STM studies of quasi-one-dimensional or layered chain structures exhibiting CDWs will be included in Sect. 6.4.

3. Complex layered materials such as the novel high-T_c oxides. Section 6.5 will be devoted to topographic STM studies of the high-T_c oxides performed at room temperature whereas their superconducting properties, as probed by low-temperature STM, are described in Chap. 8.

There also exist intercalation compounds for all three of the above categories. It is well known [6.1] that the quasi-two-dimensional character of layered materials can be gradually weakened or enforced by intercalation. In particular, the electronic structure and the elastic properties can be changed quite significantly and therefore, intercalation can be expected to influence the STM response. STM studies of the intercalation compounds of graphite will be discussed in Sect. 6.2.

In Sect. 6.1, we describe STM investigations of the simplest, one-component layered material: graphite. The wealth of unexpected experimental findings for this apparently simple material demonstrates the challenges in understanding STM results on layered materials in general.

6.1 STM Studies of Graphite

The most common form of graphite in nature is hexagonal graphite, for which the crystal structure is shown in Fig. 6.1. It is built up by layers with a honeycomb arrangement of carbon atoms which are strongly covalently bonded to one another. The nearest neighbor distance is 1.42 Å whereas the in-plane lattice constant is 2.46 Å. The layers are spaced 3.35 Å apart and are held together by weak van der Waals forces. Neighboring layers are shifted relative to each other leading to an ABABAB . . . stacking sequence and a c-axis lattice constant of 6.70 Å perpendicular to the layers. This stacking sequence gives rise to two non-equivalent carbon atom sites within the two-dimensional surface unit cell: one carbon atom (A-site) has a neighboring carbon atom directly below in the next layer whereas the other carbon atom (B-site) is located above the centre (hollow site) of the sixfold carbon ring in the second layer.

Since naturally occurring graphite single crystals are relatively small and difficult to obtain, the most widely studied form of graphite by STM is highly orientated pyrolytic graphite (HOPG). This polycrystalline material with a hexagonal structure has a relatively large grain size (about 3–10 μm) and a good c-axis orientation (misorientation angle less than 2°). The rhombohedral form

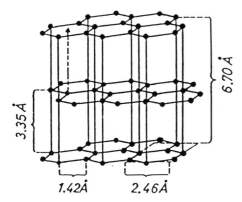

Fig. 6.1. Crystal structure of graphite. The lattice constants are 2.46 Å (in-plane) and 6.70 Å (perpendicular to the layers)

with an ABCABC ... stacking sequence usually forms 5–15% of natural graphite. However, it cannot be isolated from the hexagonal form and is therefore difficult to study experimentally.

The easy sample preparation of HOPG, by peeling off a few carbon sheets with adhesive tape, together with the inertness of the graphite surface towards chemical reactions have made it the standard test and calibration sample for atomic resolution STMs working either in air or in liquids. Despite the relatively simple structure and easy surface preparation, interpretation of the STM results for graphite is far from straightforward. In the following, we will describe some of the peculiarities of experimental STM results on the graphite surface, together with an interpretation which will provide insight into the different sources of image contrast in STM studies of layered materials.

6.1.1 Site Asymmetry, Energy-Dependent Corrugation, Tunneling Spectroscopy and Electronic Structure of the Graphite Surface

The first atomic resolution STM studies of the graphite surface in ultra-high vacuum and in air were reported in 1986 by *Binnig* et al. [6.2] and by *Park* and *Quate* [6.3], respectively. Three important conclusions were drawn from the experimental results. Firstly, the large scale STM images confirmed directly the atomic flatness of some thousand square ångstrom areas of cleaved graphite. Secondly, STM was shown to be able to distinguish between non-equivalent carbon sites and thirdly, an energy-dependent corrugation was found in qualitative agreement with theory [6.2]. In Fig. 6.2, the comparison between experimental and model data from [6.2] is reproduced. A spacing of about 2.5 Å was observed between the maxima in the STM image, indicating that only every other carbon atom appears as a protrusion which is denoted as the "carbon site asymmetry". The assignment of the measured maxima to the A-site carbon atoms, however, was not correct as pointed out in subsequently published theoretical work [6.4–6] which showed that the B-sites should give rise to a higher tunneling current compared with the A-sites. The site asymmetry has

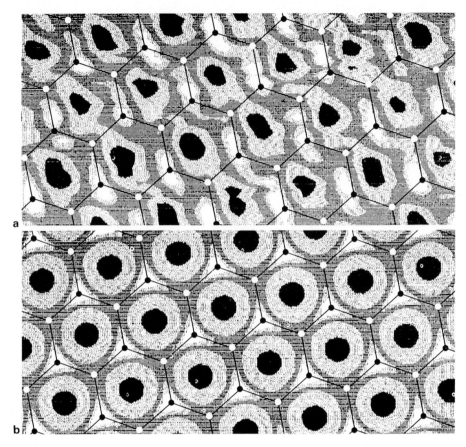

Fig. 6.2. Comparison of (**a**) experimental and (**b**) model STM data on graphite with a honeycomb overlay showing the positions of the two non-equivalent top-layer carbon sites. From [6.2]

been explained by the particular symmetry of the wave functions at the Fermi surface of graphite near the $\bar{K}$ point in the surface Brillouin zone [6.5–7]. It has also been shown [6.5,6] that the site asymmetry is nearly independent of the polarity and decreases with increasing magnitude of the bias voltage (Fig. 6.3). The observed site asymmetry on the graphite surface has neatly demonstrated that STM images generally do not reflect the surface atomic structure but are strongly influenced by the local electronic density of states [6.8]. Therefore, STM results can strongly contradict results from helium scattering experiments which probe the total charge density at the surface.

Besides the observation of the carbon site asymmetry, the second important experimental result reported in [6.2] was the strong decrease of the measured corrugation amplitude with increasing bias voltage, as demonstrated in Fig. 6.4. The corrugation amplitude of about 1 Å at a bias voltage of 50 mV is signific-antly larger than the value of 0.2 Å as determined from helium diffraction data,

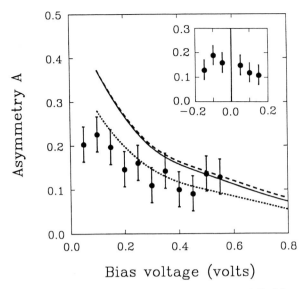

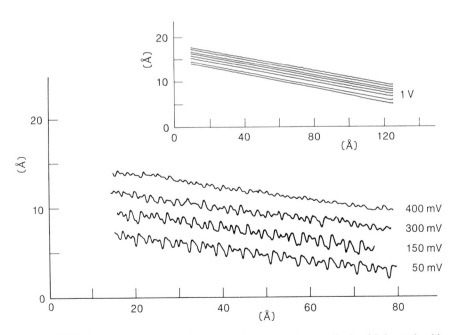

Fig. 6.3. Site asymmetry $A = [I(B) - I(A)]/[I(B) + I(A)]$ of the tunneling current I as a function of bias voltage V determined theoretically and experimentally. From [6.5]

Fig. 6.4. STM line scans showing the decrease of the corrugation amplitude with increasing bias voltage. From [6.2]

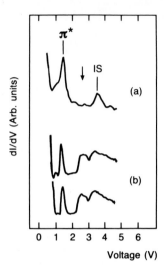

Fig. 6.5. Local tunneling spectra taken from two different locations of a HOPG surface. In spectrum (*a*) only two peaks appear whereas spectrum (*b*) shows an additional peak which is probably defect related. From [6.11]

indicating again the importance of density-of-states effects for STM imaging. The energy-dependent corrugation on the graphite surface as seen by STM was theoretically predicted [6.9] and has since been discussed in more detail [6.5,6].

Tunneling spectroscopic data on the graphite surface up to a bias voltage of 5 V have been reported by *Reihl* et al. [6.10] and by *Fuchs* and *Tosatti* [6.11]. The local tunneling spectra exhibited two and sometimes three peaks (Fig. 6.5) which were ascribed to a bulk π^*-antibonding state, an interlayer state and a possible defect-related state being only occasionally visible. The spectroscopic STM data were found to be in reasonable agreement with experimental results from inverse photoemission and with theoretical predictions [6.8].

6.1.2 Giant Corrugations, Tip–Sample Interaction and Elastic Response of the Graphite Surface

In the early work of *Binnig* et al. [6.2] it was noted that enhanced corrugation amplitudes as large as 2.5 Å were occasionally observed on the graphite surface; these could not be explained at the time. Since then, giant corrugation amplitudes of up to 8 Å [6.11] and even greater in STM studies of the graphite surface performed in air in the constant current mode of operation have been reported (Fig. 6.6). *Soler* et al. [6.12] attributed these giant corrugations to elastic deformations of the graphite surface induced by atomic forces between tip and sample surface, leading to a significant enhancement of the electronically based corrugation (Fig. 6.7). For the first time, it was realized that the STM image contrast may not always be determined by the local density of states [6.8] but can in special cases be dominated by tip–sample interaction, particularly at a small tip–surface separation. One consequence of this interaction is that the forces between tip and sample have to be taken into account whenever soft

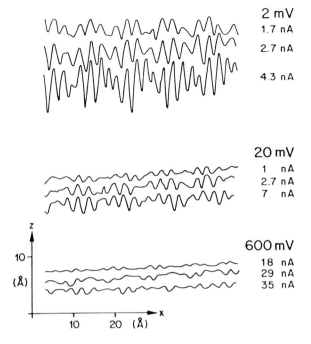

2 mV
1.7 nA
2.7 nA
4.3 nA

20 mV
1 nA
2.7 nA
7 nA

600 mV
18 nA
29 nA
35 nA

Fig. 6.6. STM line scans on the graphite surface obtained in air at atmospheric pressure. The dependence of the corrugation amplitude on the tunneling gap resistance and therefore the tip–sample separation is evident. From [6.12]

samples such as graphite and other layered materials are studied by STM at low tunneling gap resistances. For STM experiments performed in ambient air, the forces acting between the tip and the graphite surface have been measured directly [6.13, 14]. A remarkably high value of about 10^{-6} N was obtained. *Mamin* et al. [6.15] showed that for STM studies of graphite performed in air, the force interaction between tip and sample surface is mediated by a contamination layer and that the force necessary to deform the surface is not acting on an atomic scale, as was assumed by *Soler* [6.12], but must be spread over several thousand square ångstroms. *Pethica* [6.16] also argued that the actual contact area must be more than one atom wide. He emphasized that imaging a simple periodic structure does not strictly require a tunnel "beam" of single-atom width, but rather a fluctuation in total conductivity between tip and sample which varies in registry with unit lattice shear. Particularly for layered materials which can easily be sheared along the basal planes, sliding of the planes just beneath the tip may produce a periodic increase and decrease of the injected tunnel current. Such an image contrast mechanism may explain the observations by *Smith* et al. [6.17] that atomic resolution studies of the graphite surface can be performed, even with point contact between tip and sample surface, where the potential barrier has already collapsed.

It is important to realize that the theoretical concepts developed for large tip–surface separations, where both electrodes can be considered as independent [6.8], must break down at sufficiently small separations or even at point contact.

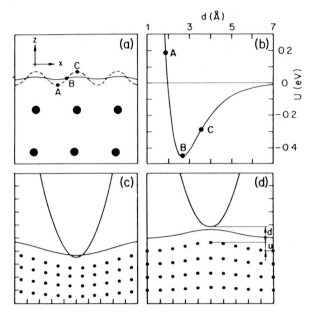

Fig. 6.7 Explanation of the giant corrugations: (**a**) The contour of constant local density of states (dashed line) which should be followed by the STM tip [6.8] differs strongly from the contour of total charge density (solid line) for the graphite surface. The points A, B, C therefore lie in different parts of the potential for the tip–surface interaction (**b**), leading to a compression of the graphite surface for the tip at point A and an expansion for the tip at point C. This results in an enhanced corrugation due to the elastic response of graphite. From [6.12]

New mechanisms for current flow might then become dominant, possibly via tip-induced localized states [6.18–21] or by quantum conductance [6.22, 23].

6.1.3 Anomalous STM Images

Apart from the commonly seen carbon site asymmetry and the occasionally observed giant corrugations, the observation of anomalous STM images of the graphite surface has further complicated the interpretation of experimental data. These anomalous STM images show triangular arrays of triangles, honeycomb arrays and triangular arrays of ellipses or linear row-like structures (Fig. 6.8).

Mizes et al. [6.24] gave an explanation for the anomalous images by assuming multiple atomic tips. They argued that STM images of graphite and other layered materials are dominated by only three independent Fourier components. A nonideal, multiple tip can change the relative amplitudes and shift the relative phases of these components, leading to changes of shape and amplitude of the maxima. Other anomalous STM images of the graphite surface (Moiré patterns) were observed near grain boundaries (Fig. 6.9). Such images were explained by assuming multiple isolated tips which scan over different

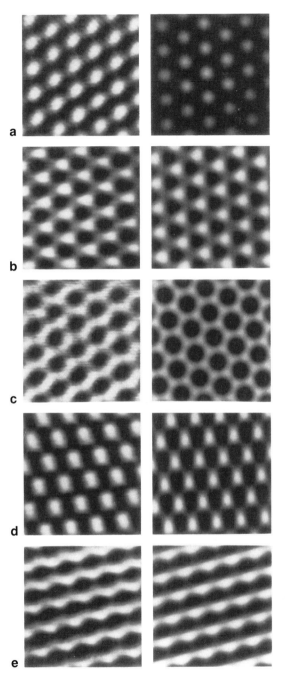

Fig. 6.8a–e. Anomalous STM images of the graphite surface obtained experimentally (left-hand column) and theoretically (right-hand column). For the computer-generated images, a linear combination of three sine waves with amplitudes and phases adjusted to match the experimental data was taken. A multiple tip is assumed to be responsible for the modification of the Fourier coefficients. From [6.24]

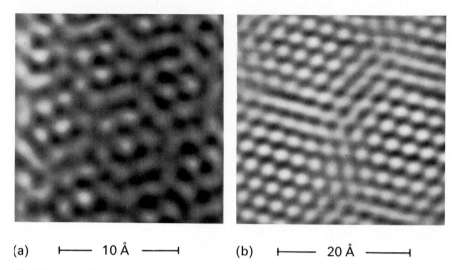

(a) ⊢——— 10 Å ———⊣   (b) ⊢——— 20 Å ———⊣

Fig 6.9a, b. Moiré patterns observed in the vicinity of grain boundaries at the graphite surface. These patterns can also be explained by multiple tip imaging. From [6.25]

grains and contribute simultaneously to the image [6.25]. Multiple isolated tips can therefore affect the image if surface deformation mediated by a contamination layer occurs, as described by *Mamin* et al. [6.15].

Kobayashi and *Tsukada* [6.26] gave another interesting explanation for anomalous STM images of the graphite surface. They showed that in the case where special tip electronic levels are active in tunneling, anomalous images such as linear-chain structures are obtained. Therefore, anomalous images are not necessarily due to multiple tip imaging.

Finally, an STM image of the graphite surface has been presented [6.27] where every carbon atom becomes equally visible (Fig. 6.10). This honeycomb lattice, which directly reflects the atomic surface structure, could only be obtained under special experimental conditions. At present, one can only speculate about the origin of this type of anomalous STM image.

6.1.4 STM Imaging of Defects

STM as a local probe is ideally suited for the investigation of nonperiodic surface structures such as point defects, steps, dislocations, grain boundaries or other extended defects. After the description and discussion of the possible sources for the STM image contrast on the ideal graphite lattice, we will now report on STM studies of defects on graphite surfaces. Such investigations can also help to clarify the STM imaging mechanism, for instance by giving estimates for the area of contact between tip and sample surface.

Bryant et al. [6.28] first reported atomic scale observations of defects on gold-sputtered graphite samples. The defects appeared as protrusions in charge

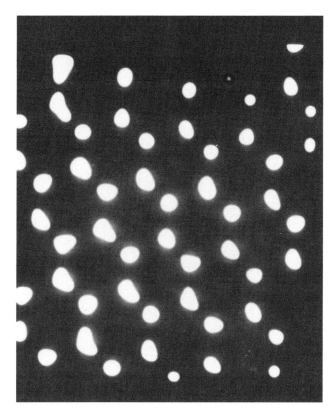

Fig. 6.10. STM image of the graphite surface where every atom becomes visible. The spacing between the tip and the sample has been reduced below a critical value. From [6.27]

density as long as the tip–sample separation exceeded a critical value. At smaller distances, however, the protrusions were not evident in STM images obtained in the variable current mode of operation. To explain their experimental results, *Bryant* et al. modeled these defects as gold atoms which lie just below the surface layer. They argued that at small tip–surface separations, the STM image contrast is dominated by the graphite electronic states. As the tip is moved away from the surface, only the slowly decaying component of the gold *s*-like state should be left. *Mizes* and *Harrison* [6.29] used the tight-binding approximation to study the effects of a substitutional impurity in graphite. They found two signatures of such a point defect. Firstly, the height of the protrusion at the defect site in the STM image is distance dependent and secondly, the shape of the protrusion may be triangular. It was also pointed out that the protrusion as seen in the STM image is not an image of the impurity but instead its effect on the graphite wave function. An assignment of chemical identity to the impurity may only be obtained by studying the distance dependence of the height of the protrusion at the defect site, which itself depends on the type of impurity. *Soto*

Fig. 6.11. This 40×40 Å^2 STM image of graphite obtained at a constant tunneling current of 1.2 nA and a bias voltage of 270 mV shows the electronically induced ($\sqrt{3} \times \sqrt{3}$) superlattice in the vicinity of an isolated defect. From [6.34]

[6.30, 31] later extended these theoretical studies by using the extended Hückel tight-binding method.

Another interesting observation can be made from STM studies of point and linear defects on the graphite surface [6.32, 33, 25]. A ($\sqrt{3} \times \sqrt{3}$) superlattice often appears in the vicinity of the defect (Fig. 6.11). This superlattice is ascribed to long-range electronic perturbations caused by the defect. It has been argued [6.32] that the local perturbation of the charge density by the defect can lead to periodic oscillations similar to Friedel oscillations. The oscillations have a wavelength $\sqrt{3}$ times that of the graphite lattice and decay away from the defect over some ten lattice constants with a power law rather than an exponential law. The symmetry of the oscillations were shown to reflect the nature of the defect (Fig. 6.12).

However, it should also be noted that atomic scale defects on carbon-bombarded graphite surfaces have been studied by STM [6.35] where neither the distance dependence reported by *Mizes* and *Harrison* [6.29] for substitutional impurities nor the ($\sqrt{3} \times \sqrt{3}$) superlattice were observed (Fig. 6.13). It is therefore likely that different types of defects on graphite surfaces may have different signatures in STM images.

6.1.5 STM Studies of Clusters on the Graphite Surface

In Sect. 6.1.4 the anomalous distance dependence observed above defects on the surface of a gold-sputtered graphite sample, where the gold atoms were assumed to be located below the surface layer [6.28], was described. In the following, STM studies of metal clusters deposited on the graphite surface will be briefly summarized.

It has been shown [6.36–41] that STM allows the investigation of the static structure and dynamic behaviour of Au, Ag, Cu and Al clusters and islands on an atomic scale. The adsorption sites of monomers, dimers and trimers, and the internal structure of extended monolayer islands have been determined [6.36, 38–40]. The metal islands often contain ordered regions separated by grain boundaries, while the atomic arrangement at the periphery is disordered. The ordered parts exhibit rectangular lattices which are not close packed as in the bulk fcc structure and are incommensurate with the graphite substrate lattice (Fig. 6.14).

Ganz et al. [6.40] also studied the appearance of silver dimers and trimers as a function of tip–surface separation. They found that the silver clusters become invisible at larger separations if the STM is operated in the variable current mode, whereas in the constant current mode of operation the clusters appear larger with increasing separation. Although the authors presented a qualitative argument for this behaviour, it should be noted that the distance dependence observed in the variable current mode is opposite to the one reported by *Bryant* et al. [6.28] for the gold-sputtered graphite sample. If the discrepancy between

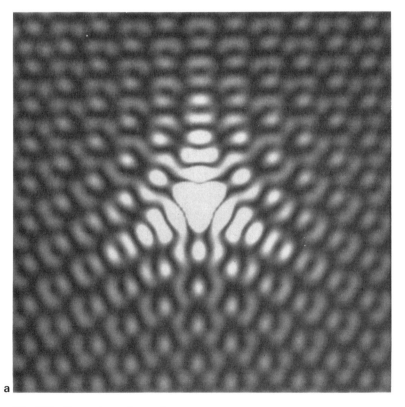

a

Fig. 6.12a. See next page for caption

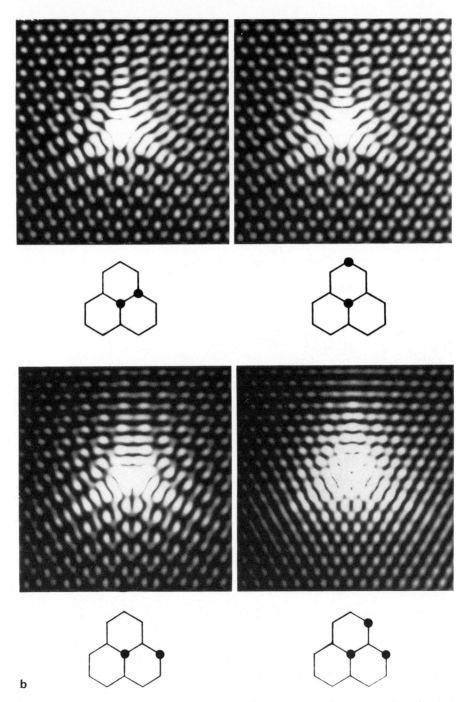

Fig. 6.12. Computer simulation of a single (**a**) and four different multiple point defects (**b**) on a graphite surface. The symmetry of the superlattice reflects the symmetry of the defect. From [6.32]

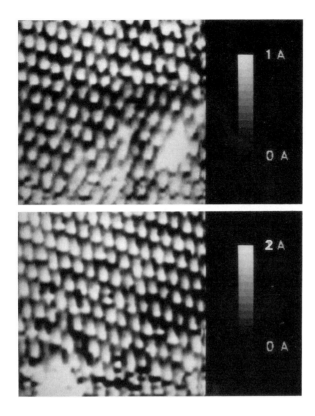

Fig. 6.13. STM images of atomic scale defects on a carbon-bombarded graphite surface without the $(\sqrt{3} \times \sqrt{3})$ superlattice. The tunneling parameters were $I = 0.5\,\text{nA}$ and $U = 60\,\text{mV}$. From [6.35]

the two experimental results could be attributed to the location of the metal species – either on top or beneath the graphite surface layer – one would be able to distinguish between these two locations by distance-dependent STM measurements. More theoretical and experimental work is necessary to clarify this point.

6.2 STM Studies of Graphite Intercalation Compounds

Graphite intercalation compounds (GICs) [6.42, 43] have been the subject of numerous experimental and theoretical studies in recent years as they provide model systems for the investigation of 2D physics [6.44]. GICs are obtained by intercalating material into the graphite galleries (Fig. 6.15). This can be achieved by a number of different preparation techniques, such as the two-temperature gas phase method, reactions with the liquid phase intercalant or electrochemical preparation methods [6.42, 43]. Depending on the preparation conditions, different stages of GICs can be obtained where the stage number n denotes the number of carbon (C) layers between two neighboring intercalant (I) layers

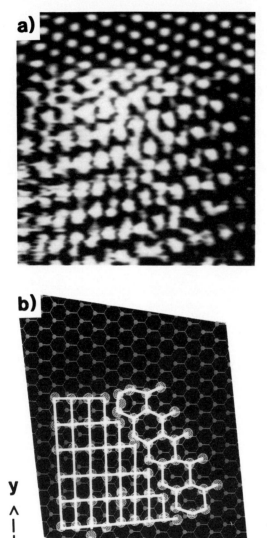

Fig. 6.14. (a) This $35 \times 35 \, \text{Å}^2$ STM current image shows a monolayer Au island on a graphite surface. **(b)** Computer model of the internal structure of the island. From [6.40]

(Fig. 6.16). The overall intercalant concentration of a GIC is primarily given by the relative amount of I layers to C layers, which is represented by the reciprocal stage number $1/n$, and – for a given stage number n – by the in-plane density of the I layer. The intercalant sublattice is known to exhibit in-plane order or disorder, depending on the intercalant species, its concentration and the investigated temperature range. Ordered I sublattices can be either commensurate or incommensurate with the C sublattice.

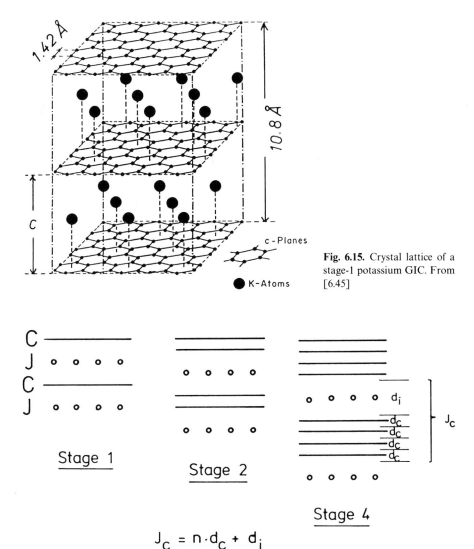

Fig. 6.15. Crystal lattice of a stage-1 potassium GIC. From [6.45]

$$J_c = n \cdot d_c + d_i$$

Fig. 6.16. Different stages of graphite intercalation compounds

The in-plane lattice constant of the graphitic host lattice is not affected much by intercalation. Only small changes (i.e. of about 1%) may occur which reflect the changes in the electronic properties due to the charge transfer upon intercalation. However, the distance between the carbon layers of 3.35 Å in pure graphite can be drastically increased in GICs and varies between 3.71 Å for Li GICs and up to 10 Å for acceptor compounds. (Depending on the direction of charge transfer between the intercalant and the graphite host, one distinguishes between donor GICs and acceptor GICs.)

Interestingly, the ABABAB ... stacking sequence of hexagonal graphite mentioned in Sect. 6.1 is generally not preserved in GICs. The same type of C layer is usually found on each side of an I layer, giving an AIAIAI ... sequence for stage 1 GICs and, for instance, an AIABABIBABAIA ... sequence for stage 4 GICs. The I layers may or may not be regularly stacked in the c-direction perpendicular to the layers.

To understand the phenomenon of stage transformations, a domain model for GICs has been proposed by *Daumas* and *Hérold* (Fig. 6.17). In this model all graphite galleries are equally filled, but not continuously, thereby forming islands of the intercalant material.

Unfortunately, a major obstacle has usually to be overcome for the investigation of GICs. Donor GICs such as alkali metal GICs easily oxidize, whereas many acceptor GICs are either easily decomposed or hygroscopic.

Nevertheless, the extension of STM studies from graphite to GICs is of interest for several reasons. Firstly, the surface structure of GICs is almost unknown, whereas the microscopic bulk structure has been studied in detail by high-resolution transmission electron microscopy. STM is ideally suited for the determination of the local atomic-scale surface structure, to study its relation to defects in the graphitic host lattice and to investigate possible domain formation. Secondly, the charge transfer between the intercalant and the graphite host leads to changes in the electronic structure of GICs compared to pure graphite. Most importantly, the local density of states at the Fermi level and the Fermi surfaces are significantly modified in GICs [6.43] (in fact, the semi-metal graphite is turned into a synthetic metal). Therefore, STM studies of GICs can complement the results obtained on pure graphite. Thirdly, as another consequence of intercalation, the elastic response of GICs should generally differ from that of pure graphite, which might be reflected in a different STM response at low tunneling gap resistances.

In the following, we will first describe STM investigations of donor and acceptor GICs before the experimental results are discussed and compared with theoretical predictions.

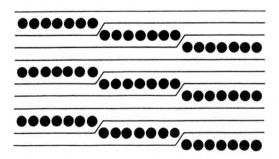

Fig. 6.17. Daumas-Hérold domain model for staging in a stage-3 GIC

6.2.1 Donor Graphite Intercalation Compounds

Among the donor compounds, alkali metal GICs are the simplest prototype systems from a conceptual point of view. However, in contrast to the graphite surface which is rather inert and can be studied by STM on an atomic scale even in air and liquids [6.46], the surfaces of alkali metal GICs are known to be highly reactive and immediately oxidized when exposed to air. Additionally, the low vapour pressure of alkali metals prevents STM studies in ultra-high vacuum at room temperature. Successful STM measurements on alkali metal GICs at room temperature can, however, be performed in an inert gas environment [6.47–53]. The microscope is operated, for instance, in a stainless steel glove box containing a high purity argon atmosphere and connected to a gas purification system which reduces impurities such as H_2O, O_2 and N_2 to below the detection limit of 1 ppm. The samples are cleaved in situ on the microscope's stage just before the STM measurements are started. Atomic resolution studies can be performed for up to several hours before the STM image quality deteriorates, most likely due to surface oxidation.

STM overview images of GICs generally show atomically flat surface regions together with highly defective areas [6.47–49, 54, 55]. The mean terrace size is significantly smaller and the density of defects higher compared with pure graphite. This observation is most likely explained by the creation of faults in the graphite host lattice during intercalation. On the atomic scale, a variety of surface structures have been observed on alkali metal GICs which will now be described beginning with the lightest alkali metal, Li GIC. STM results on stage-1 compounds with ordered in-plane structures of the intercalant at room temperature will be presented first and the results on stage-2 compounds will then be summarized.

The first atomic resolution STM studies of the surface of a stage-1 Li GIC (C_6Li) by *Anselmetti* et al. [6.47–49] have revealed three different types of triangular superlattices with in-plane lattice constants of $3.5 \pm 0.2\,\text{Å}$, $4.2 \pm 0.2\,\text{Å}$ and $4.9 \pm 0.2\,\text{Å}$ (Fig. 6.18). The lattice constant of $4.2\,\text{Å}$ corresponds to a commensurate ($\sqrt{3} \times \sqrt{3}$) superlattice (Fig. 6.19a) which can also be found in the bulk of C_6Li, whereas the lattice constant of $4.9\,\text{Å}$ indicates the presence of a commensurate (2×2) superlattice (Fig. 6.19b) which is usually found in the bulk of stage-1 heavy alkali metal GICs. The third observed lattice constant of $3.5\,\text{Å}$ could be attributed to an incommensurate superlattice of nearly close-packed lithium. It is known that pure hexagonal close-packed lithium has a lattice constant of $3.1\,\text{Å}$. Electrostatic repulsion, originating from an almost complete charge transfer from the lithium to the graphitic layers in C_6Li, is likely to increase this spacing. However, a definite assignment of this superlattice with the smallest in-plane lattice constant is difficult as long as the underlying graphitic host lattice cannot be imaged simultaneously. Unfortunately, the non-observation of the underlying host lattice in STM images of the C_6Li surface seems to be typical, in contrast to STM images of the heavier alkali

Fig. 6.18. (a–c) Three constant current STM images showing the different types of triangular superlattices which were observed on the surface of C_6Li. The imaged surface area (18×18 Å^2) is the same for all three figures. The tunneling parameters were $I = 1$ nA and $U = -100$ mV. From [6.47]

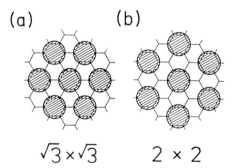

Fig. 6.19. Bulk structures of **(a)** C_6M and **(b)** C_8M graphite intercalation compounds where M denotes an alkali metal

metal GIC surfaces which are described below. The observation of the super-lattice itself was independent of the applied bias voltage whereas the measured STM corrugation showed a bias dependence. At small bias voltage (< 200 mV), the corrugation was found to be similar to pure graphite (about 1 Å) whereas at larger bias voltage, a strong decrease of the corrugation was observed [6.47]. A remarkable superlattice on a scale of several hundred ångstroms (Fig. 6.20) can occasionally be seen on the surface of C_6Li. The origin of this superlattice is still unclear but nevertheless its observation emphasizes the potential of STM for revealing novel surface structures also on a mesoscopic scale.

STM images of the surface of stage-1 K GICs (C_8K) [6.51] show a commensurate (2×2) superlattice, as found in the bulk of C_8K and also the

Fig. 6.20. STM image of the C_6Li surface showing a triangular superstructure on a mesoscopic scale. The lattice constant is about 300 Å, the height of the individual structures about 20 Å and their average diameter about 60 Å. The tunneling parameters were $I = 1$ nA and $U = 400$ mV. From [6.34]

Fig. 6.21. STM image of the C_8K surface recorded at a tunneling current of 4 nA and a sample bias voltage of -15 mV. The white bar corresponds to 10 Å. From [6.51]

underlying graphitic host lattice (Fig. 6.21). However, one often observes only a graphitic surface structure without any superlattices superimposed [6.48, 49], in contrast to the C_6Li surface. A bias dependence of the appearance of the (2×2) superlattice was not found [6.51].

On the surfaces of the heavier alkali metal stage-1 GICs, C_8Rb and C_8Cs, commensurate (2×2) superlattices and the underlying graphitic host lattice were observed (Fig. 6.22) [6.50, 52, 53]. However, in contrast to the observations on the C_6Li and C_8K surfaces, one also finds non-hexagonal superlattices on the surfaces of stage-1 Rb and Cs GICs [6.52, 53]. In Fig. 6.23 an STM image of the

a   b

Fig. 6.22. STM images of (**a**) $(50 \times 60\ \text{Å}^2)$ and (**b**) $(25 \times 30\ \text{Å}^2)$ surface areas of C_8Rb showing the (2×2) superlattice with the underlying graphitic host lattice. The tunneling parameters were $I = 3.4$ nA and $U = 50$ mV. From [6.52]

Fig. 6.23. STM image $(85 \times 85\ \text{Å}^2)$ of a C_8Cs surface showing the one-dimensional $(\sqrt{3} \times 4)$ superlattice with the simultaneously imaged underlying graphitic host lattice. The regularity of the superlattice structure is disturbed by a link between two neighboring chains. The tunneling parameters were $I = 2.3$ nA and $U = 6.4$ mV. From [6.53]

C_8Cs surface is presented which reveals a one-dimensional ($\sqrt{3} \times 4$) superlattice superimposed on the underlying graphitic host lattice. A superlattice defect can also be seen in this image. Similar non-hexagonal superlattices, ($\sqrt{3} \times 2$) and ($\sqrt{3} \times \sqrt{13}$), have been found in the bulk of unsaturated Cs GICs by using scanning transmission electron microscopy [6.56]. Therefore, a dilution of the intercalant at the surface might be responsible for the observed one-dimensional structures. Besides the ($\sqrt{3} \times 4$) superlattice, a variety of other one-dimensional superlattices were observed on the surfaces of C_8Rb and C_8Cs. One example is shown in Fig. 6.24. The one-dimensional superlattices obviously break the threefold symmetry of the underlying graphitic host lattice. This symmetry

Fig. 6.24. Another one-dimensional super-lattice structure observed at the C_8Cs surface. The large interchain distance is $4a_0$ and the smaller one is $3.5a_0$, where $a_0 = 2.46$ Å is the in-plane lattice constant of the underlying graphitic host lattice. The tunneling parameters were $I = 2.6$ nA and $U = 43$ mV. From [6.53]

Fig. 6.25. STM image (100×90 Å^2) of the C_8Rb surface showing the three possible rotational domains of the one-dimensional ($\sqrt{3} \times 4$) superlattice structure. The tunneling parameters were $I = 2$ nA and $U = 20$ mV. From [6.52]

breaking is, however, restricted to a local microscopic scale. On a more macroscopic scale, the threefold symmetry can be regained by the occurrence of three rotational domains of the one-dimensional superlattices. This can indeed be observed by STM, as shown in Fig. 6.25. The measured corrugation on the C_8Rb and C_8Cs surfaces is 1.5 ± 0.5 Å for the (2×2) superlattice, 0.5 ± 0.3 Å for the $(\sqrt{3} \times 4)$ superlattice and 1.0 ± 0.3 Å for the graphitic host lattice [6.52]. The bias voltage, which was varied in the range $\pm (20-400 \text{ mV})$ in the experiments by *Anselmetti* et al. [6.52, 53] and in the range $\pm (750 \text{ mV})$ in the experiments by *Kelty* and *Lieber* [6.50], did not influence the observed superlattice structures.

STM studies of stage-2 K, Rb and Cs GICs revealed the absence of a superlattice structure at the surface [6.50]. Instead, the images appear similar to those obtained on pure graphite with a well-pronounced carbon site asymmetry. However, the measured corrugation of 0.4 ± 0.2 Å on the stage-2 alkali metal GICs at a small bias voltage of 15 mV is significantly lower than on pure graphite.

6.2.2 Acceptor Graphite Intercalation Compounds

Acceptor GICs were first studied by *Gauthier* et al. [6.54, 55]. These experiments were performed at room temperature in a vacuum of 10^{-6} Torr. A graphitic surface structure was observed on stage-1 $FeCl_3$ GICs and on bromine compounds. The measured corrugation was in both cases comparable with that on pure graphite.

In a more recent STM study in ambient air of an acceptor compound, stage-1 $CuCl_2$ GIC, an interesting bias dependence was found [6.57]. At positive sample bias voltage ($+10 \text{ mV}$), the honeycomb lattice of the graphite surface was observed whereas at negative sample bias voltage (-10 mV), a monoclinic superlattice was found which was ascribed to the intercalate $CuCl_2$ layer lying just below the top graphitic surface layer. For the graphitic honeycomb lattice observed at positive sample bias, the measured corrugation amplitude was found to be reduced by more than 60% compared with pure graphite.

Spectroscopic STM data on another acceptor compound, stage-1 $CoCl_2$ GIC, was reported by *Tanaka* et al. [6.58]. They found a shift in the Fermi level position of about 0.4 eV compared with pure graphite (Fig. 6.26a). This shift in the Fermi level towards lower energy is expected for acceptor-type compounds, as illustrated in Fig. 6.26b.

6.2.3 Interpretation and Comparison with Theoretical Predictions

The discussion of experimental STM results on pure graphite (Sect. 6.1) has shown that the images do not reflect the surface atomic structure but are dominated by electronic structure effects, and at low tunneling gap resistances additionally by the elastic response of the sample. The special surface electronic structure of graphite leads to the observed carbon site asymmetry and to a relatively large corrugation amplitude of about 1 Å, compared with the value of

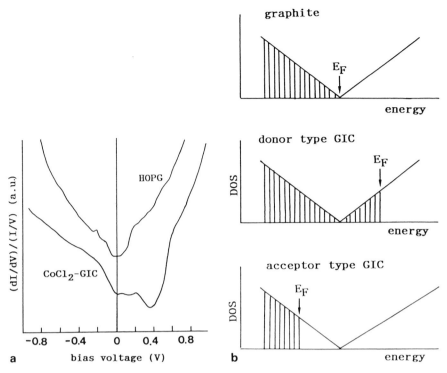

Fig. 6.26. (a) Normalized differential conductance versus bias voltage for a CoCl$_2$ GIC and HOPG measured at room temperature in ambient air. (b) Comparison of the Fermi level position in idealized pure graphite, donor GICs and acceptor GICs. From [6.58]

0.2 Å obtained from helium scattering. The elastic response of graphite can additionally enhance this corrugation, leading to giant corrugations of up to several ångstroms or more. Intercalating either donors or acceptors into graphite drastically changes its electronic structure and elastic properties. Therefore, a different STM response can be expected for GICs. In the following, the consequences of the altered surface electronic structure will first be discussed before the influence of the elastic response is taken into account.

The first theoretical predictions for STM studies on stage-1 Li GICs were made by *Selloni* et al. [6.59]. For their calculations, using empirical pseudo-potentials and a plane wave representation, they modeled the sample by a slab of four carbon and three lithium layers in a repeated slab geometry. An AIAIAIA stacking of the layers with a top carbon layer and a triangular ($\sqrt{3} \times \sqrt{3}$) superlattice of the intercalated lithium were assumed. Unlike graphite, the surface atoms of C$_6$Li are all equivalent due to the AIAI . . . stacking. Therefore, a carbon site asymmetry is not expected. The corrugation amplitude near zero bias was predicted to be reduced to 0.2–0.4 Å for C$_6$Li compared with 1.0–1.5 Å for pure graphite. A rapid decrease of the corrugation on C$_6$Li with increasing positive sample bias voltage should occur due to tunneling into the smooth empty interlayer state.

The experimentally obtained STM images of C_6Li [6.47–49] were dominated by the different superlattices and showed no sign of the underlying graphitic host lattice. The measured corrugation at small bias voltage (< 200 mV) was comparable to that found on pure graphite whereas a rapid decrease of the corrugation was observed at higher bias voltages. Only this latter observation is in agreement with the theoretical predictions. The experimental results indicate that the two assumptions made by Selloni were too restrictive. Firstly, the in-plane ordering of the alkali metal should not be kept fixed. The STM images of the C_6Li surface proved the existence of at least three different types of superlattices. Secondly, from an experimental point of view, it seems obvious that after cleaving stage-1 GICs, part of the intercalant will be left on top of the surface. For symmetry reasons, the intercalant is likely to be present on the surfaces of both parts of the cleaved sample and a rearrangement of the intercalant on top of the surface might occur. This could perhaps explain the dominance of the superlattices in the STM images and the larger corrugation at small bias voltage compared with the theoretical predictions.

Other calculations of STM images of GICs, based on tight-binding theory, were made by *Qin* and *Kirczenow* [6.60, 61]. They predicted that the corrugation amplitude and carbon site asymmetry should be sensitive to the number of graphitic layers covering the first intercalant layer, to the amount of transferred charge and its distribution and also to the surface sub-band structure. In particular, it was predicted that the STM image of an alkali metal GIC should be nearly featureless on the atomic scale if there is only one graphitic layer covering the first intercalant layer. This prediction is contradictory to all atomic resolution STM results on stage-1 alkali metal GICs obtained so far [6.47–53], leading to the conclusion that either the top surface layer is not a carbon but an alkali metal layer, as already discussed, or that there are other important electronic contributions such as possible surface driven charge density waves [6.50, 51, 62, 63]. A decrease of the measured corrugation on stage-2 alkali metal GICs [6.50] or on stage-1 compounds showing a graphitic surface structure [6.48, 49, 57] is, however, observed and is consistent with the expansion of the Fermi surface that occurs upon intercalation. *Qin* and *Kirczenow* [6.60, 61] further predicted that the carbon site asymmetry should be absent for small numbers of graphitic layers covering the first guest layer, even for AB stacking. Although this prediction is in agreement with experimental results on a stage-1 acceptor compound at positive sample bias voltage [6.57], it contradicts observations on alkali metal compounds [6.47–53]. In particular, STM images of stage-2 alkali metal GICs [6.50] showed the centered hexagonal structure with a strong carbon site asymmetry. Atomic resolution STM images of stage-1 alkali metal GICs, showing the superlattice and the underlying graphitic host lattice simultaneously [6.50–53], also reveal a pronounced carbon site asymmetry in the graphitic sublattice. This pronounced carbon site asymmetry therefore awaits an explanation.

Another important issue for understanding STM results on GICs is the distance dependence which has already been found to be of importance in STM studies of a gold-sputtered graphite surface [6.28] and a graphite surface covered with a monolayer island of silver [6.40].

Besides electronic structure effects, one should also consider the possible elastic response of the GICs. Because of the insensitivity of STM results on stage-1 alkali metal GICs to changes in the tunneling current and the applied bias voltage in the constant current mode of operation, *Kelty* and *Lieber* [6.51] concluded that local sample deformations cannot explain the observed super-lattices. However, the development of a theory for the local distortions of the graphitic planes in the vicinity of intercalated species has just been initiated [6.64] and will be important for a quantitative comparison with the measured STM corrugations.

Finally, the observed one-dimensional superlattices on the surfaces of stage-1 Rb and Cs GICs [6.52, 53], which locally break the threefold symmetry of the underlying graphitic host lattice, and the observed superstructure on a meso-scopic scale (Fig. 6.20) are waiting for a theoretical treatment. Even more unexpected superlattice structures may be discovered by STM in the near future and will significantly contribute to the field of GICs.

6.3 STM Studies of Transition Metal Dichalcogenides

Transition metal dichalcogenides (TMDs) [6.65] with layered structure are built up by a repetition of a three-layer sandwich consisting of a top chalcogenide layer (S, Se or Te), a middle transition metal layer (e.g. Ti, V, Nb, Mo, Ta, W, ...) and a bottom chalcogenide layer (Fig. 6.27a). The bonding within each sandwich is covalent, while neighboring sandwiches are held together mainly by weak van der Waals forces which facilitate cleavage. The layers usually form a triangular lattice. The coordination around the transition metal atoms can either be trigonal prismatic or octahedral (Fig. 6.27b). Accordingly, one can distinguish between the 1 T phases of TMDs (pure octahedral coordination), the 2H phases (pure trigonal prismatic coordination) and the 4Hb phases (mixed octahedral and trigonal prismatic coordination), to mention only a few. In some TMDs such as WTe_2, transition metal atoms are, however, displaced from the centre of the coordination units leading to zig-zag and slightly buckled chains of metal atoms. Consequently, the chalcogenide sheets become slightly corrugated to accommodate the shifts of the metal atoms (Fig. 6.28).

Electrically, the TMDs cover a wide spectrum from insulators (e.g. HfS_2) through semiconductors (e.g. MoS_2, $MoTe_2$, WSe_2, SnS_2) and semimetals (e.g. WTe_2) to true metals (e.g. NbS_2). The surfaces of TMDs, consisting of chalco-genide layers, are relatively inert, as is graphite. Therefore, atomic resolution STM studies can be performed even in ambient air or liquids.

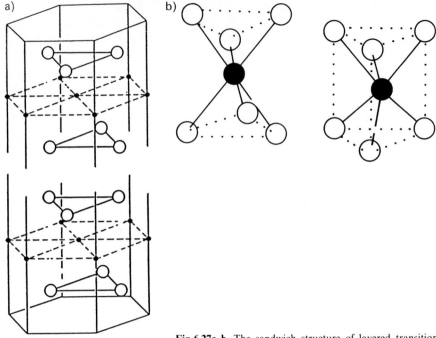

O S or Se atoms

• Ti or Ta atoms

Fig 6.27a, b. The sandwich structure of layered transition metal dichalcogenides consists of a top chalcogenide layer (S, Se or Te), a middle transition metal layer and a bottom chalcogenide layer. (**b**) Trigonal prismatic (left hand) and octahedral (right hand) coordination units

STM images of TMDs such as NbSe$_2$ [6.66–71], MoS$_2$ [6.72–76], WSe$_2$ [6.75, 77] and MoTe$_2$ [6.78] all show a triangular lattice (Fig. 6.29). Since both the top chalcogenide layer and the second transition metal layer of these materials are built up by a triangular lattice with the same in-plane lattice constant, it cannot be decided a priori whether the top chalcogenide or the second metal layer is imaged by STM. From the experience obtained in interpreting STM images of non-layered semiconductors and metals, one would expect to see the top surface chalcogenide layer [6.66–68, 71] whereas a faint, secondary triangular lattice which is sometimes additionally observed [6.70, 73] could be assigned to the second metal layer. However, it has been argued [6.72, 77] that the metal d_{z^2} orbitals, which protrude vertically to the surface, may provide the major contribution to the tunneling current and that the STM image contrast is dominated by the second transition metal layer. Finally, it has been proposed on the basis of a scanning tunneling spectroscopic study [6.79] that the chalcogenide layer is imaged at negative sample bias voltage whereas the transition metal layer is imaged at positive sample bias.

To decide experimentally which layer of the TMDs is actually probed by STM, *Tang* et al. [6.78] have studied WTe$_2$. In this material, as already

WTe2

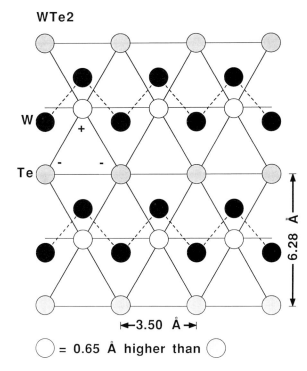

W

Te

|←— 3.50 Å —→|

◯ = 0.65 Å higher than ◯

6.28 Å

Fig. 6.28. Surface structure of WTe$_2$. The top Te layer is slightly buckled, with alternate rows of atoms 0.65 Å higher than the others. The W layer is comprised of zig-zag rows of atoms below the Te layer. From [6.80]

Fig. 6.29. STM image of MoS$_2$ showing a triangular lattice with an in-plane lattice constant of 3.2 Å. From [6.137]

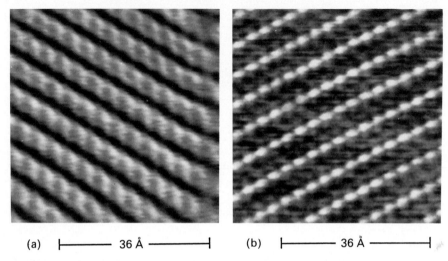

(a) ├────── 36 Å ──────┤   (b) ├────── 36 Å ──────┤

Fig. 6.30a, b. Different STM images of WTe$_2$ obtained with the same tunneling parameters. Image (a) apparently shows the zig-zag rows of W atoms. Image (b) shows only one bright spot per unit cell suggesting that image (a) is obtained with a double tip. This image (b) can well be explained by imaging the top Te layer. From [6.80]

mentioned, a distorted coordination of the metal atoms leads to the formation of zig-zag and slightly buckled chains of metal atoms, thereby making the chalcogenide and transition metal layers non-equivalent. The STM images of WTe$_2$ showed zig-zag patterns which were attributed to the tungsten rows. Tang concluded that their results therefore unambiguously demonstrate that it is possible for the STM to image sub-surface atoms. However, in a more recent STM study of WTe$_2$ by *Albrecht* [6.80], it is shown that the zig-zag pattern obtained by Tang is most likely explained by anomalous STM imaging due to a multiple tip (Fig. 6.30). Such anomalous STM imaging has already been discussed in Sect. 6.1 for graphite. Apart from multiple tip imaging, one can only contemplate that different tip electronic states make the dominant contribution to the tunneling current leading to the different STM images of WTe$_2$ presented in Fig. 6.30. In both cases, however, it has to be concluded that on the basis of the currently available experimental data, it cannot be decided which layer of the TMDs is indeed imaged by STM.

Finally, it should be noted that, similar to the observations in STM studies of the graphite surface performed in air at low tunneling gap resistances, giant corrugations of up to 8 Å have been found in STM studies of TMDs in air [6.68, 70]. These giant corrugations were again attributed to elastic deformations of the sample due to tip–surface forces which are likely to be transmitted by a surface contamination layer.

6.4 STM Studies of Charge Density Waves

Some transition metal dichalcogenides (TMDs) and other quasi-low-dimensional materials exhibit electronic phase transitions from a normal into a condensed state which is called the charge density wave (CDW) state [6.81, 82]. For the case of a one-dimensional metal, *Peierls* [6.83] has shown that the system of conduction electrons coupled to the underlying periodic lattice is not stable at low temperatures. The ground state of the coupled electron–phonon system is characterized by a gap in the single particle excitation spectrum and by a collective mode formed by electron–hole pairs involving the wave vector $2k_F$, where k_F is the Fermi wave vector. The spatial modulation of the concentration of conduction electrons associated with this collective mode can be regarded as a standing wave of charge density. The novel condensed state is therefore denoted as the charge density wave state. The charge density modulation is accompanied by a periodic lattice distortion since ionic displacements are necessary to screen the electronic modulation. For an arbitrary band filling, the spatial period of the charge density modulation and the accompanying periodic lattice distortion is incommensurate with the underlying atomic lattice.

In real materials, CDW formation is favoured if Fermi surface nesting occurs because the electron–hole interaction at $2k_F$ can then drive the Fermi surface instability. This is more likely to be the case for quasi-one- and two-dimensional metals such as some of the TMDs. The amplitude of the charge density modulation is dependent on the detailed geometry of the Fermi surface, which itself is dependent on the type of material and even on the specific phase (e.g. 1 T, 2 H or 4 Hb for the TMDs). The transition from the normal to the CDW state can occur in a wide temperature range (e.g. from above 600 K to about 20 K for TMDs). Additionally, incommensurate (I) to commensurate (C) transitions can be observed within the CDW state. For some TMDs, these I–C transitions occur well below the CDW transition temperature or are not observed in the experimentally accessible temperature range. Since the CDW formation is directly connected with a modification of the conduction electron wavefunctions and the local density of states at the Fermi level, STM has become an important tool for the investigation of the local aspects of CDW formation.

6.4.1 Charge Density Waves in Transition Metal Dichalcogenides

The first STM observation of CDWs on cleaved surfaces of 1 T-TaS$_2$ at 77 K was reported by *Coleman* et al. [6.84]. The STM images showed a triangular lattice with a CDW lattice spacing of $(3.5 \pm 0.3) a_0$, where a_0 denotes the atomic lattice constant of 1 T-TaS$_2$ (Fig. 6.31). This triangular lattice is generated by three CDWs at relative orientations of 120°. Since the first observation, a large number of TMDs exhibiting CDWs have been investigated by STM in a temperature range from 300 K down to 4 K [6.85, 86]. The most intensively

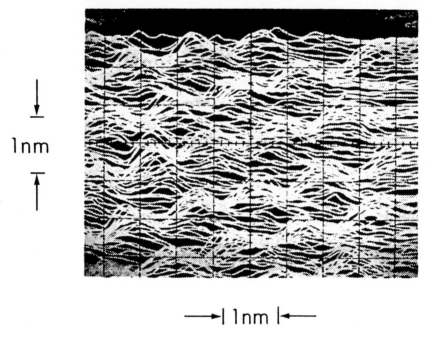

1 nm

⟶| 1 nm |⟵

Fig. 6.31. The first STM observation of charge density waves on a cleaved surface of 1T-TaS$_2$ obtained at 77 K. The tunneling parameters were $I = 5.5$ nA and $U = 50$ mV. From [6.84]

studied materials are the 1 T phases of TaS$_2$ and TaSe$_2$ [6.84–94] which show a CDW transition at a temperature of about 600 K well above room temperature. While the CDW of 1 T-TaSe$_2$ becomes commensurate at 473 K, the CDW state of 1 T-TaS$_2$ exhibits several intermediate phases before becoming commensurate at 150 K. The 1 T phases of TaS$_2$ and TaSe$_2$ are the only TMDs which allow the investigation of the CDW state even at room temperature. For both materials, it is possible to image the CDW superlattice and the underlying atomic lattice simultaneously by STM (Fig. 6.32). Therefore, the orientational relationship between these two lattices can be determined and the commensurability can be directly checked.

A large corrugation of 2–4 Å for the CDW superlattice is measured by STM whereas the corrugation for the underlying atomic lattice usually remains below 1 Å. This anomalous high corrugation of the CDW superlattice in the 1 T phases of TaS$_2$ and TaSe$_2$ may have several origins. Firstly, a large electron condensation into the CDW state can lead to a strong CDW amplitude. Secondly, the CDW modulation can additionally be enhanced by several possible mechanisms. *Tersoff* [6, 7] pointed out that the anomalous corrugations observed on the 1 T phases of TaS$_2$ and TaSe$_2$ may have the same origin as for graphite. Whenever the Fermi surface collapses to a point at the corner of the surface Brillouin zone, which is the case for graphite and for 1 T-TaS$_2$ and 1 T-TaSe$_2$, the

Fig. 6.32. STM image (140×140 Å^2) of 1T
TaSe$_2$ showing the commensurate ($\sqrt{13}$
$\times \sqrt{13}$) CDW superlattice which is superimposed on the atomic lattice. The
tunneling parameters were $I = 2.8$ nA and
$U = -80$ mV. From [6.34]

STM image will correspond to an individual state. The nodal structure of this
state gives rise to a large corrugation with the periodicity of the unit cell. The
electronically enhanced corrugation may further be enlarged by tip–sample
interaction forces, as first proposed by *Soler* et al. [6.12] and discussed in
Sect. 6.1.

Besides the anomalous size of the measured STM corrugations for the CDW
superlattice on the 1T phases of TaS$_2$ and TaSe$_2$, an anomalous distance
dependence of the corrugation has been observed. The corrugation of the CDW
superlattice was shown to progressively decrease with increasing current and
therefore with decreasing tip–surface separation [6.88]. This is in contrast to the
distance dependence of the corrugation for the atomic lattice where an increase
of the corrugation with decreasing tip–surface distance can usually be observed.
A possible explanation for this anomalous distance dependence was given by
Coleman et al. [6.88] on the basis of *Tersoff's* theory [6.7]. However, this theory
would only explain a distance independence of the corrugation, and not a
decrease of the corrugation with decreasing tip–surface separation. Another
explanation for the anomalous distance dependence was given by *Meyer* et al.
[6.95, 96] on the basis of a comparative STM and atomic force microscopy
(AFM) study of the 1T phases of TaS$_2$ and TaSe$_2$. Due to increasing tip–sample
interaction forces at decreasing tip–surface separation, a suppression of the
CDW state might occur, caused by the local applied pressure (a suppression of
the CDW state in the 1T phases of TaS$_2$ and TaSe$_2$ with increasing applied
pressure is well known from earlier experiments on a macroscopic scale). This
explanation would then imply that the mean diameter of the spot, over which
the pressure is locally applied, is of the order of the coherence length of the
CDW state.

Large corrugations of up to 4 Å have also been measured in STM studies of
the 1T phases of TiS$_2$ and TiSe$_2$ [6.97] where only the latter material exhibits a

CDW state below 202 K. The anomalous corrugations on these materials can again be explained by electronic structure effects [6.7] and by an additional enhancement due to tip–surface interaction forces because the compressibility perpendicular to the layers is even larger than on the corresponding Ta-based compounds.

The 2 H phases of TaS_2, $TaSe_2$ and $NbSe_2$ also show CDW superlattices in STM images at temperatures lower than the corresponding CDW transition temperatures (75 K, 122 K and 33 K respectively). However, the measured corrugations of the CDW superlattices are significantly smaller compared to the 1 T phases of TaS_2 and $TaSe_2$, indicating a smaller charge modulation of the conduction electrons [6.87–89].

The 4 Hb phases are particularly interesting since they are composed of alternating sandwiches of octahedral and trigonal prismatic coordination. Independent CDW formation can be observed in the two different types of sandwiches with wave vectors similar to those of the pure single coordination phases but with slightly reduced transition temperatures. STM studies of the 4 Hb phases of TaS_2 and $TaSe_2$ [6.87, 89, 98–100] indeed revealed two types of image depending on which sandwich is located on top of the surface. STM images of the sandwich with octahedral coordination show the strong ($\sqrt{13} \times \sqrt{13}$) CDW superlattice with an anomalous large corrugation, as found for the pure 1 T phases. On the other hand, STM images of the sandwich with trigonal prismatic coordination reveal either the absence of a CDW superlattice at room temperature and at 77 K, or an extremely weak (3 × 3) CDW superlattice at 4 K which is characteristic for the pure 2 H phases. Occasionally, a superposition with the strong ($\sqrt{13} \times \sqrt{13}$) CDW superlattice of the 1T sandwich beneath can be observed [6.98] as shown in Fig. 6.33.

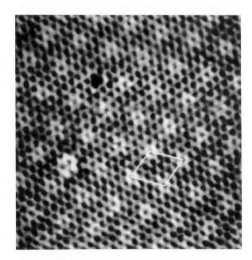

Fig. 6.33. Room temperature STM image (70×70 Å^2) of the trigonal prismatic co-ordinated sandwich of 4Hb-TaS_2 showing the atomic lattice with a lattice defect. A weak ($\sqrt{13} \times \sqrt{13}$) superlattice is superimposed which is attributed to the CDW state in the underlying octahedral coordinated sandwich. The tunneling parameters were $I = 2$ nA and $U = 60$ mV. From [6.34]

Finally, it should be noted that, as in the case of superconductors (Chap. 8), STM can be used to determine the CDW energy gap on a local scale [6.99–101]. Spatially resolved STM measurements of this CDW gap in the vicinity of CDW defects and domain boundaries (Sect. 6.4.2) will be particularly interesting.

6.4.2 Charge Density Wave Defects and Domains

In the 1970s and 1980s, the CDW state was intensively investigated by bulk-sensitive methods such as neutron, electron and X-ray diffraction [6.81], which are sensitive to the periodic lattice distortion, and by electrical transport and susceptibility measurements, which are directly sensitive to the changes in the Fermi surface due to the CDW formation. Helium scattering experiments showed that CDWs propagate up to the topmost atom layer of a crystal [6.102, 103]. While all these experimental techniques provide information averaged over a macroscopic sample volume or surface area, STM as a local probe allows the direct investigation of CDW superlattice defects and CDW domain structures. These inhomogeneities in the CDW superlattice can be correlated directly with the structure of the underlying atomic lattice which can simultaneously be imaged by STM. Besides being a local probe, STM offers the advantage of directly sensing the local density of states at the Fermi level, which allows detailed investigation of the distribution of the local conduction electron density at CDW defects or domain boundaries.

The first observation of a point defect in the CDW superlattice was reported by *Giambattista* et al. [6.89]. An isolated charge maximum in the CDW superlattice of 1 T-TaSe$_2$ was completely quenched, tentatively attributed to a Ta-atom vacancy in the subsurface transition metal layer. The perturbation of the CDW maximum appeared remarkably localized since the surrounding charge-density contours revealed very little distortion. An STM image showing an isolated defect and a defect complex in the CDW superlattice of 1 T-TaSe$_2$, together with the underlying atomic lattice, is presented in Fig. 6.34. However, even in cases where the underlying atomic lattice can be imaged simultaneously, it proves difficult to correlate the defect structures in the CDW superlattice with possible atomic lattice defects – vacancies or impurities – in the sub-surface transition metal layer.

To study the relationship between CDW superlattice defects and atomic lattice impurities in more detail, *Wu* et al. [6.105–108] substitutionally doped their samples of 1 T-TaS$_2$ and 1 T-TaSe$_2$ with Ti or Nb, thereby obtaining materials of the general form $I_y Ta_{1-y} X_2$ (I = Ti, Nb; X = S, Se). Several important conclusions were drawn from STM studies of these doped TMDs.

Firstly, in the case of Ti doping, the average CDW wavelength was found to increase with increasing dopant concentration [6.105], which directly proves that the CDW responds on average to the decreased size of the Fermi surface of the doped samples (substituting Ti for Ta leads to a decrease of the conduction electron density by one electron per Ti atom). In contrast, the CDW wavelength was found to be unchanged in the case of Nb doping [6.107] where the

Fig. 6.34. STM image (120×120 Å^2) of 1T TaSe$_2$ obtained at 77 K. An isolated defect and a defect complex can be observed in the commensurate ($\sqrt{13} \times \sqrt{13}$) CDW super-lattice together with the simultaneously imaged underlying atomic lattice. The tunneling parameters were $I = 3$ nA and $U = -224$ mV. From [6.104]

substitution of Nb for Ta does not significantly perturb the size of the Fermi surface. Intercalation of lithium into 1 T-TaS$_2$, on the other hand, leads to electron donation to the Ta d-band, and therefore to an increase of the Fermi surface. As a consequence, a reduction of the CDW wavelength is observed [6.109].

Secondly, *Wu* et al. showed that the corrugation amplitudes of the CDW superlattice decreases with increasing dopant concentration. This can be explained within Tersoff's theory [6.7, 105].

Thirdly, the local CDW structure was found to become significantly distorted in response to the random lattice potential associated with the impurities [6.105, 108]. It was found that the frequency of localized CDW defects increases linearly with dopant concentration (Fig. 6.35) and that several dopant centers may be necessary to cause a single CDW defect [6.108]. At low dopant concentration, the localized defects consist of a CDW amplitude distortion, while for higher concentrations the defects consist of a coupled amplitude-phase distortion. CDW twin domains that nucleate at these amplitude-phase defects were directly imaged (Fig. 6.36).

A different type of CDW domains can be observed by STM in pure 1 T-TaS$_2$. This material exhibits a nearly commensurate (NC) CDW phase between the high temperature incommensurate (I) phase, which exists above 350 K, and the low temperature commensurate (C) phase, which exists below 200 K. Upon warming, another nearly commensurate triclinic (T) phase is observed in the temperature range between 220 K and 280 K before entering the NC phase. Two different models have been proposed for nearly commensurate phases. In the first model [6.110] the CDW amplitude and phase are uniform, whereas in the second model [6.111–113] a hexagonal domain-like structure of the CDW superlattice is predicted. Within the domains, the CDWs are commensurate

Fig. 6.35a–d. STM images ($220 \times 220 \, \text{Å}^2$) or $Ti_y Ta_{1-y} Se_2$ samples ($x = 0$, 0.02, 0.04 and 0.07 for image **a**, **b**, **c** and **d** respectively), demonstrating that localized CDW defects occur more frequently with increasing dopant concentration. The tunneling parameters were $I = 2$ nA and $U = 10$ mV. From [6.108]

with the atomic lattice, while at the domain boundaries the CDW phase changes and the CDW amplitude decreases.

Early STM studies of the room temperature NC phase [6.90, 91, 105] did not reveal the existence of domain-like CDW structures, possibly because of the relatively small scan areas (typically below $100 \times 100 \, \text{Å}^2$). More recently obtained STM images on a larger scale, however, show a hexagonal domain structure with diffuse domain walls [6.93, 94]. Within a given domain, the rotation angle of the CDW superlattice relative to the atomic lattice is close to the commensurate value of 13.9°, independent of temperature. On the other hand, the hexagonal domain structure is rotated by about 6° relative to the

CDW superlattice in a single domain. This rotation is opposite to the direction of rotation of the CDW superlattice relative to the atomic lattice. The domain period increases linearly from about 60 Å to about 90 Å between 350 K and 230 K with a clear discontinuity at the first-order I–NC transition at 350 K. The diffuse domain walls are 2–3 CDW periods wide (ca. 24–36 Å) and nearly independent of temperature. In these domain wall regions, the CDW amplitude decreases by 0.5–1 Å. A STM image showing the hexagonal domain structure at 300 K is presented in Fig. 6.37.

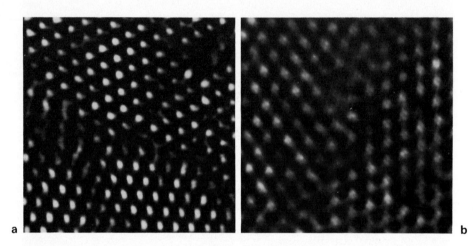

a   b

Fig. 6.36. (a) A 165×165 Å^2 image of a $Ti_{0.07}Ta_{0.93}Se_2$ sample that exhibits a CDW twin domain. (b) A 140×140 Å^2 image of another CDW twin domain observed in a $Ti_{0.08}Ta_{0.92}Se_2$ sample. From [6.108]

Fig. 6.37. STM image (200×200 Å^2) of 1T TaS_2 obtained in the NC phase at 300 K. A triangular domain-like structure of the CDW superlattice is observed with a period of about 70 Å. The tunneling parameters were $I = 6$ nA and $U = -19$ mV. From [6.104]

Wu et al. also studied the effect of doping on the hexagonal domain-like CDW structure in the NC phase of 1 T-TaS$_2$. Upon Nb doping, a distortion of the hexagonal domains was observed [6.107] which can be explained by the influence of the random lattice potential introduced by the Nb impurities. On the other hand, Ti doping leads to a destruction of the hexagonal domain structure [6.105]. This might be explained by the fact that the hexagonal domain structure is favored by the electrostatic interaction between the lattice and commensurate CDW domains, and the magnitude of this interaction is proportional to the CDW amplitude. Since Ti doping leads to a uniformly low CDW amplitude, as already discussed, interactions with the random lattice potential lead to the disordered CDW structure without domains.

Finally, it should be noted that in the nearly commensurate triclinic (T) phase of 1 T-TaS$_2$, CDW domain structures have also been directly observed by STM [6.90, 91]. In contrast to the circular shape of the domains observed in the NC phase (Fig. 6.37), the domains of the T phase exhibit a long narrow shape (Fig. 6.38). These stripe domains are approximately 60–70 Å across at 230 K and rotated by an angle of about 26° relative to the CDW superlattice.

The STM studies of defects and domains of the CDW superlattice have impressively demonstrated the potential of STM to reveal local inhomogeneities in the distribution of charge from the atomic to the mesoscopic scale.

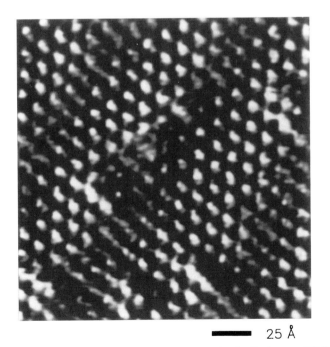

■ 25 Å

Fig. 6.38. STM image of 1T-TaS$_2$ obtained in the T phase at 225 K. The CDW superlattice is clearly observed in long narrow domains separated by walls where the CDW amplitude is suppressed. The tunneling parameters were $I = 5$ nA and $U = +20$ mV. From [6.90]

6.4.3 Charge Density Waves in Quasi-One-Dimensional Systems

Besides layered materials with a quasi-two-dimensional crystal and electronic band structure, compounds with a quasi-one-dimensional band structure can also show CDW formation at low temperatures. The transition metal trichalcogenides (TMTs) such as $NbSe_3$ and TaS_3 belong to this class of quasi-1D compounds. They are composed of linear chains built up by trigonal prisms stacked along the chain axis. The quasi-1D electronic band structure results from a strong overlap of the transition metal d-orbitals along the chain direction with no direct d–d overlap perpendicular to the chains. The microscopic linear chain structure of the TMTs is reflected in the macroscopic shape of the crystals which form thin fibrous ribbons or needles. Cleavage of these crystals is more difficult than for the layered-type compounds. STM studies of the TMTs $NbSe_3$ and TaS_3 from room temperature down to 4 K clearly reveal the linear chain structure of these materials [6.69, 92, 114–119]. STM images of $NbSe_3$ obtained at 4 K additionally show two independent CDWs, where the CDW modulation is observed to be substantially localized on different chains for the separate CDWs [6.115]. The CDW structure of orthorhombic TaS_3 has also been studied by STM [6.116, 118, 119].

Another group of quasi-1D compounds, the bronze materials such as $K_{0.3}MoO_3$ ("blue bronze") and $Rb_{0.3}MoO_3$, also exhibit CDWs. In these compounds, the MoO_6 octahedra form chains (Fig. 6.39), separated by chains of alkali metal atoms, leading to a quasi-two-dimensional crystal structure but to an almost one-dimensional electronic band structure. Room temperature STM studies of $K_{0.3}MoO_3$ [6.120, 121] and $Rb_{0.3}MoO_3$ [6.122] revealed the chains of MoO_6 octahedra. In addition, single and multiple defects in blue bronze were

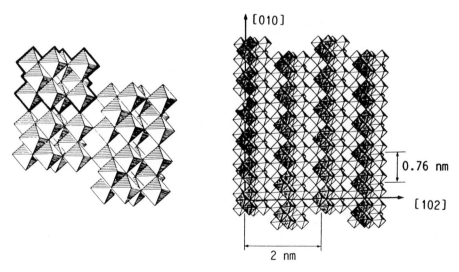

Fig. 6.39. The crystal structure of $K_{0.3}MoO_3$ consists of corner- and edge-sharing MoO_6 octahedra forming chains in the [010] direction and a $(\bar{2}01)$ cleavage plane.

Fig. 6.40. Room temperature STM image $(200 \times 200 \ \text{Å}^2)$ of $K_{0.3}MoO_3$ showing the chains of MoO_6 octahedra together with single and multiple defects. The tunneling parameters were $I = 2.6$ nA and $U = -135$ mV. From [6.121]

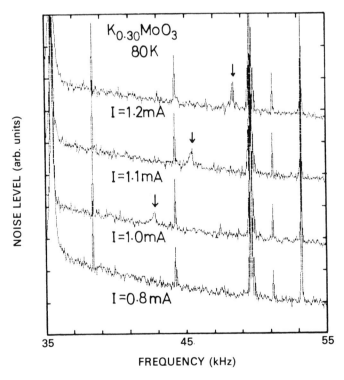

Fig. 6.41. Frequency spectra of the tunneling current flowing between a fixed STM tip and a blue bronze sample under a sample bias current at a temperature of 80 K. If the bias current exceeds a threshold value, a new peak (indicated by arrows) appears in the frequency spectra of the tunneling current. From [6.124]

characterized by STM [6.121], as shown in Fig. 6.40. The defects play an important role in CDW materials since, in contrast to superconductors, the phase of the CDW condensate can be pinned to the lattice through the interaction with impurities, lattice imperfections, grain boundaries and other defect structures. However, in contrast to the quasi-2D compounds, the pinning energy per electron in the quasi-1D compounds is small enough that the dynamics of the collective CDW mode can be observed in response to finite amplitude dc or ac excitations [6.82].

The first report of a STM observation of sliding CDWs in blue bronze was by *Nomura* and *Ichimura* [6.123, 124]. Holding the tunneling tip at a fixed position, they found a sharp peak in the frequency spectra of the tunneling current if the applied sample bias current exceeded the threshold for the CDW depinning (Fig. 6.41). The additional peak was attributed to the modulation of the tunneling gap due to the sliding motion of the CDW. The sliding velocity at the surface was found to be considerably higher than in the bulk.

In the future it will be most interesting to correlate STM observations of atomic scale defect structures with STM observations of the CDW dynamics.

6.5 STM Studies of High-T_c Superconductors

The discovery of high-T_c superconductivity in layered oxides [6.125] has led to an increasing interest in layered materials of complex structure which have hitherto been relatively neglected in solid state physics. Almost all known experimental techniques have been applied to these materials in order to investigate the mechanism of superconductivity. STM is a promising tool to investigate the relationship between properties of the superconducting state and the atomic and electronic structure on a local scale [6.126]. Firstly, STM can be used as a spectroscopic tool at low temperatures to determine the superconducting energy gap locally, which is advantageous particularly for sintered materials where the grain size hardly exceeds several microns. Secondly, the atomic and electronic structure of high-T_c superconductors (HTSCs) can be studied by STM from room temperature down to low temperatures.

In this section, we will concentrate on room temperature STM studies of the atomic and electronic structure of HTSCs with layered structure, whereas low temperature STM investigations of superconductors in general will be described in Chap. 8.

Among the different families of HTSCs, the Bi- and Tl-based copper oxides are best suited for surface investigations because clean surfaces can easily be obtained by cleaving the samples, as for other layered materials of less complex structure. The crystal structure of a Bi-based copper oxide is shown in Fig. 6.42. The cleavage process most likely leaves a BiO plane (Bi compounds) or a TlO plane (Tl compounds) on top of the surface. Atomic resolution STM investigations of Bi-based compounds seem to require either ultra-high vacuum [6.127,

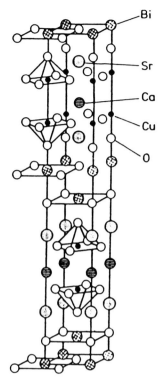

Bi₂Sr₂CaCu₂O₈

Fig. 6.42. Crystal structure of $Bi_2Sr_2CaCu_2O_{8+x}$

128] or an inert gas environment [6.34], whereas Tl-based compounds can be studied on an atomic scale even in air [6.129].

STM images of Bi-based copper oxides show the atomic lattice and an incommensurate superstructure consisting of a sinusoidal modulation with a periodicity of nine to ten unit cells (Fig. 6.43). This superstructure has also been observed by LEED and by transmission electron microscopy, and its origin has recently been explained [6.130]. The measured distance between the atomic scale protrusions on surfaces of Bi-based compounds suggests that only one species, either the Bi or the O atoms, are imaged by STM. The assignment is, however, controversial [6.127, 128]. Spectroscopic STM measurements on Bi-based compounds have revealed a very small electronic density of states at the Fermi level. This implies that the surface BiO layer is not metallic [6.128, 131, 132]. Therefore, a model for the Bi-based copper oxides was proposed consisting of an alternating stacking of metallic CuO_2 planes and nonmetallic BiO planes [6.131, 132].

STM studies of Tl-based copper oxides [6.129] also reveal the atomic lattice and a weak one-dimensional superlattice with a period of about 10 Å (Fig. 6.44). In contrast to the Bi-based compounds, both the Tl and the O sites of the top

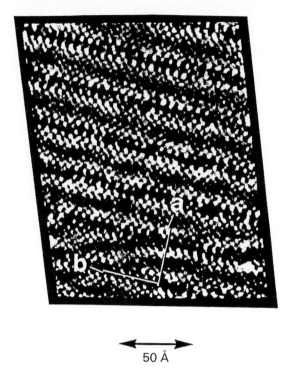

50 Å

Fig. 6.43. STM image (150 × 200 Å^2) of a $Bi_2Sr_2CaCu_2O_{8+x}$ single crystal showing a sinusoidal modulation superimposed on the atomic lattice. The tunneling parameters were $I = 0.2$ nA and $U = 150$ mV. From [6.127]

Fig. 6.44. STM image of a $Tl_2Ba_2CaCu_2O_{8+x}$ single crystal showing a one-dimensional superlattice modulation together with the underlying atomic lattice. The tunneling parameters were $I = 0.2$ nA and $U = -200$ mV. From [6.129]

surface plane can be imaged by STM, indicating that Tl and O make similar contributions to the density of states near the Fermi level.

Other families of HTSCs, such as the (123) compounds, are much more difficult to study on the atomic scale by STM because the samples cannot easily be cleaved. However, a few atomic resolution STM investigations have been reported [6.133–135]. Simultaneously recorded topographic and local conductance images of (123) compounds, fractured in ultra-high vacuum, revealed an alternation of metallic and semiconducting characteristics [6.136]. The metallic strips were attributed to parts of the surface consisting of Cu and O, whereas the semiconducting strips were ascribed to parts of the surface built up by Ba and O. Possible implications for the basic understanding of HTSCs have been discussed [6.136].

However, more combined topographic and spectroscopic STM studies from room temperature down to low temperatures are necessary to obtain an insight into the mechanism of high-T_c superconductivity in layered oxides.

6.6 Concluding Comments

This chapter has focussed on the application of STM to materials with a layered structure. Although layered materials have already been extensively studied by STM, it has become clear that the interpretation of such STM data remains a subtle art, as electronic structure effects and tip–sample interaction must be taken into account. On the other hand, the application of STM to layered materials has proven to be instructive in many ways, e.g. for the realization of the possible influence of forces in STM investigations. The experimental results described in this chapter have also demonstrated the great potential of STM for studying the local surface properties of layered materials which are almost inaccessible by other surface analytical techniques. Among these local surface properties are defects in the atomic lattice or in the superlattices of intercalated species, or charge density waves and domains of novel superlattices on surfaces of graphite intercalation compounds and high-T_c superconductors. Atomic scale structure investigations together with energy and spatially resolved measurements of the electronic density of states undoubtedly help to clarify the nature of the variety of local surface structures which can be found on layered materials.

We gratefully acknowledge all those who have provided illustrations of their work for this chapter.

References

6.1   For a comprehensive overview of layered materials see: *Physics and Chemistry of Materials with Layered Structures*, Vol. 1–6 (Reidel, Dordrecht)
6.2   G. Binnig, H. Fuchs, Ch. Gerber, H. Rohrer, E. Stoll, E. Tosatti: Europhys. Lett. **1**, 31 (1986)

6.3   Sang-il Park, C.F. Quate: Appl. Phys. Lett. **48**, 112 (1986)
6.4   I.P. Batra, N. Garcia, H. Rohrer, H. Salemink, E. Stoll, S. Ciraci: Surf. Sci. **181**, 126 (1987)
6.5   D. Tománek, S.G. Louie, H.J. Mamin, D.W. Abraham, R.E. Thomson, E. Ganz, J. Clarke: Phys. Rev. B **35**, 7790 (1987)
6.6   D. Tománek, S.G. Louie: Phys. Rev. B**37**, 8327 (1988)
6.7   J. Tersoff: Phys. Rev. Lett. **57**, 440 (1986)
6.8   J. Tersoff, D.R. Hamann: Phys. Rev. B**31**, 805 (1985)
6.9   A. Selloni, P. Carnevali, E. Tosatti, C.D. Chen: Phys. Rev. B**31**, 2602 (1985)
6.10   B. Reihl, J.K. Gimzewski, J.M. Nicholls, E. Tosatti: Phys. Rev. B**33**, 5770 (1986)
6.11   H. Fuchs, E. Tosatti: Europhys. Lett. **3**, 745 (1987)
6.12   J.M. Soler, A.M. Baro, N. Garcia, H. Rohrer: Phys. Rev. Lett. **57**, 444 (1986)
6.13   S.L. Tang, J. Bokor, R.H. Storz: Appl. Phys. Lett. **52**, 188 (1988)
6.14   C.M. Mate, R. Erlandsson, G.M. McClelland, S. Chiang: Surf. Sci. **208**, 473 (1989)
6.15   H.J. Mamin, E. Ganz, D.W. Abraham, R.E. Thomson, J. Clarke: Phys. Rev. B**34**, 9015 (1986)
6.16   J.B. Pethica: Phys. Rev. Lett. **57**, 3235 (1986)
6.17   D.P.E. Smith, G. Binnig, C.F. Quate: Appl. Phys. Lett. **49**, 1166 (1986)
6.18   S. Ciraci, I.P. Batra: Phys. Rev. B**36**, 6194 (1987)
6.19   E. Tekman, S. Ciraci: Phys. Scripta **38**, 486 (1988)
6.20   E. Tekman, S. Ciraci: Phys. Rev. B**40**, 10286 (1989)
6.21   S. Ciraci, A. Baratoff, I.P. Batra: Phys. Rev. B**41**, 2763 (1990)
6.22   S. Ciraci, E. Tekman: Phys. Rev. B**40**, 11969 (1989)
6.23   C.J. Chen: Mod. Phys. Lett. B**5**, 107 (1991)
6.24   H.A. Mizes, Sang-il Park, W.A. Harrison: Phys. Rev. B**36**, 4491 (1987)
6.25   T.R. Albrecht, H.A. Mizes, J. Nogami, Sang-il Park, C.F. Quate: Appl. Phys. Lett. **52**, 362 (1988)
6.26   K. Kobayashi, M. Tsukada: J. Vac. Sci. Technol. A**8**, 170 (1990)
6.27   G. Binnig: Specul. Sci. Technol. **10**, 345 (1987)
6.28   A. Bryant, D.P.E. Smith, G. Binnig, W.A. Harrison, C.F. Quate: Appl. Phys. Lett. **49**, 936 (1986)
6.29   H.A. Mizes, W.A. Harrison: J. Vac. Sci. Technol. A**6**, 300 (1988)
6.30   M.R. Soto: J. Microsc. **152**, 779 (1988)
6.31   M.R. Soto: Surf. Sci. **225**, 190 (1990)
6.32   H.A. Mizes, J.S. Foster: Science **244**, 559 (1989)
6.33   J.P. Rabe, M. Sano, D. Batchelder, A.A. Kalatchev: J. Microsc. **152**, 573 (1988)
6.34   D. Anselmetti: "Rastertunnelmikroskopie an Synthetischen Metallen"; Ph.D. Thesis, University of Basel (1990)
6.35   R. Coratger, A. Claverie, F. Ajustron, J. Beauvillain: Surf. Sci. **227**, 7 (1990)
6.36   D.W. Abraham, K. Sattler, E. Ganz, H.J. Mamin, R.E. Thomson, J. Clarke: Appl. Phys. Lett. **49**, 853 (1986)
6.37   A.M. Baró, A. Bartolome, L. Vazquez, N. Garcia, R. Reifenberger, E. Choi, R.P. Andres: Appl. Phys. Lett. **51**, 1594 (1987)
6.38   E. Ganz, K. Sattler, J. Clarke: Phys. Rev. Lett. **60**, 1856 (1988)
6.39   E. Ganz, K. Sattler, J. Clarke: J. Vac. Sci. Technol. A**6**, 419 (1988)
6.40   E. Ganz, K. Sattler, J. Clarke: Surf. Sci. **219**, 33 (1989)
6.41   A. Humbert, R. Pierrisnard, S. Sangay, C. Chapon, C.R. Henry, C. Claeys: Europhys. Lett. **10**, 533 (1989)
6.42   M.S. Dresselhaus, G. Dresselhaus: Adv. Phys. **30**, 139 (1981)
6.43   P. Pfluger, H.-J. Güntherodt: In *Festkörperprobleme* (Advances in Solid State Physics) Vol. XXI, ed. by J. Treusch (Vieweg, Braunschweig 1981) pp. 271
6.44   M.S. Dresselhaus: Synth. Met. **12**, 5 (1985)
6.45   W. Rüdorff: Adv. Inorg. Chem. **1**, 223 (1959)
6.46   J. Schneir, R. Sonnenfeld, P.K. Hansma, J. Tersoff: Phys. Rev. B**34**, 4979 (1986)
6.47   D. Anselmetti, R. Wiesendanger, H.-J. Güntherodt: Phys. Rev. B**39**, 11135 (1989)

6.48   D. Anselmetti, R. Wiesendanger, V. Geiser, H.R. Hidber, H.-J. Güntherodt: J. Microsc. **152,** 509 (1988)
6.49   R. Wiesendanger, D. Anselmetti, V. Geiser, H.R. Hidber, H.-J. Güntherodt: Synth. Met. **34,** 175 (1989)
6.50   S.P. Kelty, C.M. Lieber: Phys. Rev. **B40,** 5856 (1989)
6.51   S.P. Kelty, C.M. Lieber: J. Phys. Chem. **93,** 5983 (1989)
6.52   D. Anselmetti, V. Geiser, G. Overney, R. Wiesendanger, H.-J. Güntherodt: Phys. Rev. **B42,** 1848 (1990)
6.53   D. Anselmetti, V. Geiser, D. Brodbeck, G. Overney, R. Wiesendanger, H.-J. Güntherodt: Synth. Met. **38,** 157 (1990)
6.54   S. Gauthier, S. Rousset, J. Klein, W. Sacks, M. Belin: J. Vac. Sci. Technol. **A6,** 360 (1988)
6.55   S. Gauthier, S. Rousset, J. Klein, W. Sacks, M. Belin: In *The Structure of Surfaces II*, ed. by J.F. van der Veen, M.A. Van Hove, Springer Ser. Surf. Sci., Vol. 11 (Springer, Berlin, Heidelberg 1988) pp. 71
6.56   D.M. Hwang, N.W. Parker, M. Utlaut, A.V. Crewe: Phys. Rev. **B27,** 1458 (1983)
6.57   C.H. Olk, J. Heremans, M.S. Dresselhaus, J.S. Speck, J.T. Nicholls: Phys. Rev. **B42,** 7524 (1990)
6.58   M. Tanaka, W. Mizutani, T. Nakashizu, N. Morita, S. Yamazaki, H. Bando, M. Ono, K. Kajimura: J. Microsc. **152,** 183 (1988)
6.59   A. Selloni, C.D. Chen, E. Tosatti: Phys. Scripta **38,** 297 (1988)
6.60   X. Qin, G. Kirczenow: Phys. Rev. **B39,** 6245 (1989)
6.61   X. Qin, G. Kirczenow: Phys. Rev. **B41,** 4976 (1990)
6.62   M. Lagues, J.E. Fischer, D. Marchand, C. Fretigny: Solid State Commun. **67,** 1011 (1988)
6.63   D. Marchand, C. Fretigny, N. Lecomte, M. Lagues: Synth. Met. **23,** 165 (1988)
6.64   D. Tomanek, G. Overney, H. Miyazaki, S.D. Mahanti, H.-J. Güntherodt: Phys. Rev. Lett. **63,** 876 (1989)
6.65   J.A. Wilson, A.D. Yoffe: Adv. Phys. **18,** 193 (1969)
6.66   H. Tokumoto, H. Bando, W. Mizutani, M. Okano, M. Ono, H. Murakami, S. Okayama, Y. Ono, K. Watanabe, S. Wakiyama, F. Sakai, K. Endo, K. Kajimura: Jpn. J. Appl. Phys. **25,** L621 (1986)
6.67   K. Kajimura, H. Bando, K. Endo, W. Mizutani, H. Murakami, M. Okano, S. Okayama, M. Ono, Y. Ono, H. Tokumoto, F. Sakai, K. Watanabe, S. Wakiyama: Surf. Sci. **181,** 165 (1987)
6.68   H. Bando, H. Tokumoto, W. Mizutani, K. Watanabe, M. Okano, M. Ono, H. Murakami, S. Okayama, Y. Ono, S. Wakiyama, F. Sakai, K. Endo, K. Kajimura: Jpn. J. Appl. Phys. **26,** L41 (1987)
6.69   R.V. Coleman, B. Giambattista, A. Johnson, W.W. McNairy, G. Slough, P.K. Hansma, B. Drake: J. Vac. Sci. Technol. **A6,** 338 (1988).
6.70   D.C. Dahn, M.O. Watanabe, B.L. Blackford, M.H. Jericho: J. Appl. Phys. **63,** 315 (1988)
6.71   M. Watanabe, D.C. Dahn, B. Blackford, M.H. Jericho: J. Microsc. **152,** 175 (1988)
6.72   G.W. Stupian, M.S. Leung: Appl. Phys. Lett. **51,** 1560 (1987)
6.73   M. Weimer, J. Kramar, C. Bai, J.D. Baldeschwieler: Phys. Rev. **B37,** 4292 (1988)
6.74   D. Sarid, T.D. Henson, N.R. Armstrong, L.S. Bell: Appl. Phys. Lett. **52,** 2252 (1988)
6.75   T.D. Henson, D. Sarid, L.S. Bell: J. Microsc. **152,** 467 (1988)
6.76   T. Ichinokawa, T. Ichinose, M. Tohyama, H. Itoh: J. Vac. Sci. Technol. **A8,** 500 (1990)
6.77   S. Akari, M. Stachel, H. Birk, E. Schreck, M. Lux, K. Dransfeld: J. Microsc. **152,** 521 (1988)
6.78   S.L. Tang, R.V. Kasowski, B.A. Parkinson: Phys. Rev. **B39,** 9987 (1989)
6.79   H. Bando, N. Morita, H. Tokumoto, W. Mizutani, K. Watanabe, A. Homma, S. Wakiyama, M. Shigeno, K. Endo, K. Kajimura: J. Vac. Sci. Technol. **A6,** 344 (1988)
6.80   T.R. Albrecht: "Advances in Atomic Force Microscopy and Scanning Tunneling Microscopy": Ph.D. Thesis, Stanford University (1989)
6.81   J.A. Wilson, F.J. DiSalvo, S. Mahajan: Adv. Phys. **24,** 117 (1975)
6.82   G. Grüner: Rev. Mod. Phys. **60,** 1129 (1988)

6.83   R.E. Peierls: *Quantum Theory of Solids* (Oxford University Press, Oxford 1955) pp. 108
6.84   R.V. Coleman, B. Drake, P.K. Hansma, G. Slough: Phys. Rev. Lett. **55,** 394 (1985)
6.85   R.V. Coleman, B. Giambattista, P.K. Hansma, A. Johnson, W.W. McNairy, C.G. Slough: Adv. Phys. **37,** 559 (1988)
6.86   R.V. Coleman, B. Drake, B. Giambattista, A. Johnson, P.K. Hansma, W.W. McNairy, G. Slough: Phys. Scripta **38,** 235 (1988)
6.87   C.G. Slough, W.W. McNairy, R.V. Coleman, B. Drake, P.K. Hansma, Phys. Rev. **B34,** 994 (1986)
6.88   R.V. Coleman, W.W. McNairy, C.G. Slough, P.K. Hansma, B. Drake: Surf. Sci. **181,** 112 (1987)
6.89   B. Giambattista, A. Johnson, R.V. Coleman, B. Drake, P.K. Hansma: Phys. Rev. **B37,** 2741 (1988)
6.90   R.E. Thomson, U. Walter, E. Ganz, J. Clarke, A. Zettl, P. Rauch, F.J. DiSalvo: Phys. Rev. **B38,** 10734 (1988)
6.91   R.E. Thomson, U. Walter, E. Ganz, P. Rauch, A. Zettl, J. Clarke: J. Microsc. **152,** 771 (1988)
6.92   G. Gammie, S. Skala, J.S. Hubacek, R. Brockenbrough, W.G. Lyons, J.R. Tucker, J.W. Lyding: J. Microsc. **152,** 497 (1988)
6.93   X.L. Wu, C.M. Lieber: Science **243,** 1703 (1989)
6.94   X.L. Wu, C.M. Lieber: Phys. Rev. Lett. **64,** 1150 (1990)
6.95   E. Meyer, D. Anselmetti, R. Wiesendanger, H.-J. Güntherodt, F. Lévy, H. Berger: Europhys. Lett. **9,** 695 (1989)
6.96   E. Meyer, R. Wiesendanger, D. Anselmetti, H.R. Hidber, H.-J. Güntherodt, F. Lévy, H. Berger: J. Vac. Sci. Technol. **A8,** 495 (1990)
6.97   C.G. Slough, B. Giambattista, A. Johnson, W.W. McNairy, C. Wang, R.V. Coleman: Phys. Rev. **B37,** 6571 (1988)
6.98   B. Giambattista, A. Johnson, W.W. McNairy, C.G. Slough, R.V. Coleman: Phys. Rev. **B38,** 3545 (1988)
6.99   M. Tanaka, W. Mizutani, T. Nakashizu, N. Morita, S. Yamazaki, H. Bando, M. Ono, K. Kajimura: J. Microsc. **152,** 183 (1988)
6.100 M. Tanaka, W. Mizutani, T. Nakashizu, S. Yamazaki, H. Tokumoto, H. Bando, M. Ono, K. Kajimura: Jpn. J. Appl. Phys. **28,** 473 (1989)
6.101 C. Wang, B. Giambattista, C.G. Slough, R. Coleman: Bull. Am. Phys. Soc. **35,** 209 (1990)
6.102 G.Boato, P. Cantini, R. Colella: Phys. Rev. Lett. **42,** 1635 (1979)
6.103 P. Cantini, G. Boato, R. Colella: Physica **B99,** 59 (1980)
6.104 S. Baumgartner: "Rastertunnelmikroskopie auf Ladungsdichtewellen"; Diploma Thesis, University of Basel (1990)
6.105 X.L. Wu, P. Zhou, C.M. Lieber: Phys. Rev. Lett. **61,** 2604 (1988)
6.106 X.L. Wu, P. Zhou, C.M. Lieber: Nature **335,** 55 (1988)
6.107 X.L. Wu, C.M. Lieber: J. Am. Chem. Soc. **111,** 2731 (1989)
6.108 X.L. Wu, C.M. Lieber: Phys. Rev. **B41,** 1239 (1990)
6.109 X.L. Wu, C.M. Lieber: J. Am. Chem. Soc. **110,** 5200 (1988)
6.110 C.B. Scruby, P.M. Williams, G.S. Parry: Phil. Mag. **31,** 255 (1975)
6.111 K. Nakanishi, H. Shiba: J. Phys. Soc. Jpn. **43,** 1839 (1977)
6.112 K. Nakanishi, H. Takatera, Y. Yamada, H. Shiba: J. Phys. Soc. Jpn. **43,** 1509 (1977)
6.113 A. Yamamoto: Phys. Rev. **B27,** 7823 (1983)
6.114 J.W. Lyding, J.S. Hubacek, G. Gammie, S. Skala, R. Brockenbrough, J.R. Shapley, M.P. Keyes: J. Vac. Sci. Technol. **A6,** 363 (1988)
6.115 C.G. Slough, B. Giambattista, A. Johnson, W.W. McNairy, R.V. Coleman: Phys. Rev. **B39,** 5496 (1989)
6.116 C.G. Slough, R.V. Coleman: Phys. Rev. **B40,** 8042 (1989)
6.117 G. Gammie, J.S. Hubacek, S.L. Skala, R.T. Brockenbrough, J.R. Tucker, J.W. Lyding: Phys. Rev. **B40,** 9529 (1989)
6.118 G. Gammie, J.S. Hubacek, S.L. Skala, R.T. Brockenbrough, J.R. Tucker, J.W. Lyding: Phys. Rev. **B40,** 11965 (1989)

6.119 C.G. Slough, B. Giambattista, W.W. McNairy, R.V. Coleman: J. Vac. Sci. Technol. A8, 490 (1990)

6.120 J. Heil, J. Wesner, B. Lommel, W. Assmus, W. Grill: J. Appl. Phys. 65, 5220 (1989)

6.121 D. Anselmetti, R. Wiesendanger, H.-J. Güntherodt, G. Grüner: Europhys. Lett. 12, 241 (1990)

6.122 E. Garfunkel, G. Rudd, D. Novak, S. Wang, G. Ebert, M. Greenblatt, T. Gustafsson, S.H. Garofalini: Science 246, 99 (1989)

6.123 K. Nomura, K. Ichimura: Solid State Commun. 71, 149 (1989)

6.124 K. Nomura, K. Ichimura: J. Vac. Sci. Technol. A8, 504 (1990)

6.125 J.G. Bednorz, K.A. Müller: Z. Phys. B64, 189 (1986)

6.126 M.C. Gallagher, J.G. Adler: J. Vac. Sci. Technol. A8, 464 (1990)

6.127 M.D. Kirk, J. Nogami, A.A. Baski, D.B. Mitzi, A. Kapitulnik, T.H. Geballe, C.F. Quate: Science 242, 1673 (1988)

6.128 C.K. Shih, R.M. Feenstra, J.R. Kirtley, G.V. Chandrashekhar: Phys. Rev. B40, 2682 (1989)

6.129 X.L. Wu, C.M. Lieber, D.S. Ginley, R.J. Baughman: Appl. Phys. Lett. 55, 2129 (1989)

6.130 Y. Le Page, W.R. McKinnon, J.-M. Tarascon, P. Barboux: Phys. Rev. B40, 6810 (1989)

6.131 M. Tanaka, T. Takahashi, H. Katayama-Yoshida, M. Fujinami, Y. Okabe, W. Mizutani, M. Ono, K. Kajimura: Nature 339, 691 (1989)

6.132 M. Tanaka, S. Yamazaki, M. Fujinami, T. Takahashi, H. Katayama-Yoshida, W. Mizutani, K. Kajimura, M. Ono: J. Vac. Sci. Technol. A8, 475 (1990)

6.133 N.J. Zheng, U. Knipping, I.S.T. Tsong, W.T. Petuskey, J.C. Barry: J. Vac. Sci. Technol. A6, 457 (1988)

6.134 L.E.C. van de Leemput, P.J.M. van Bentum, L.W.M. Schreurs, H. van Kempen: Physica C152, 99 (1988)

6.135 L.E.C. van de Leemput, P.J.M. van Bentum, F.A.J.M. Driessen, J.W. Gerritsen, H. van Kempen, L.W.M. Schreurs, P. Bennema: J. Microsc. 152, 103 (1988)

6.136 C.J. Chen, C.C. Tsuei: Solid State Commun. 71, 33 (1989)

6.137 K. Takata, S. Hosoki, S. Hosaka, T. Tajima: Rev. Sci. Instrum. 60, 789 (1989)

7. Molecular Imaging by STM

S. Chiang

With 19 Figures

The scanning tunneling microscope (STM) has begun to prove itself as a useful tool for atomic resolution imaging of molecules on surfaces. Applications of the technique to imaging of chemisorbed molecules in an ultrahigh vacuum environment, alkane-derived molecules, liquid crystals, and polymers are discussed.

7.1 Introduction to STM of Molecules

Early in the development of the scanning tunneling microscope (STM), people began to speculate about possible uses of the instrument for imaging molecules on surfaces. The idea of imaging molecules on surfaces with atomic resolution has implications for the determination of molecular adsorption sites and the study of chemical reactions on surfaces. Early results of experiments on molecular imaging were not particularly promising, as atomic resolution was elusive. For example, chemisorbed CO molecules were not resolved on Pt(100), even though the reconstruction of the surface induced by the CO was observed [7.1, 2]. Images of copper phthalocyanine on silver films showed low resolution, which was attributed to the effect of the applied electric field of the tunneling tip on surface diffusion of the molecules [7.3]. Finally, images of a bilayer of cadmium arachidate, deposited by the Langmuir–Blodgett technique on graphite, showed molecules moving quickly in the field of view of the STM during the experiment and were surprising because electrical current appeared to flow through 54 Å of low conductivity material [7.4].

In the last three years, substantial progress has been made in the imaging of molecules on surfaces. Molecules have been immobilized on the substrate for times long enough to perform STM measurements by several different methods: (1) strong chemisorption to a metal substrate (Sect. 7.2); (2) deposition in monolayer or bilayer films by the Langmuir–Blodgett technique [7.5] (Sects. 7.3, 5); (3) self-assembly into vast planes on a relatively inert substrate, such as graphite (Sect. 7.4); or (4) physisorption onto a substrate at low temperature [7.6]. Strongly chemisorbed molecules, mostly containing aromatic rings, have been observed on metal substrates in ultrahigh vacuum. In other cases, molecules have been studied at the interface between the solid substrate and a solution containing the molecules. Atomic scale resolution has been obtained on many different molecules, even though the conductivity mechanisms and detailed electronic structure of the molecule–substrate systems are not understood. In some cases, it has been postulated that the molecular contrast originates from the variation in work function due to the adsorption of the molecules [7.7], while in others the electronic structure of hybridized molecule–substrate orbitals appears to be more important [7.8, 9]. The choice of substrate greatly influences the appearance of the molecules, not only through its influence on the combined system's electronic structure, but also because it affects the chemical bonding, binding sites, and the arrangement of the molecules on the surface.

The following systems will be discussed in this chapter: (1) Chemisorbed molecules, which have been successfully measured with high resolution on metal single crystal surfaces in ultrahigh vacuum. (2) Images of alkane molecules and their derivatives, now obtained by several groups. (3) Many liquid crystal systems have been successfully imaged on both graphite and MoS_2. (4) Some progress has also been made on imaging polymers. (5) Finally, a few other molecules which do not fit into the above categories will be discussed at the end of the chapter.

7.2 STM of Chemisorbed Molecules in Ultrahigh Vacuum

High resolution STM images of molecules on surfaces in ultrahigh vacuum (UHV), which show individual molecules and their internal structure, have recently been obtained. Three systems have been studied in detail: (1) benzene and carbon monoxide coadsorbed on Rh(111) in two different ordered overlayers, the (3×3) [7.8] and the $c(2\sqrt{3} \times 4)$rect [7.9, 10], (2) copper phthalocyanine on Cu(100) [7.11] and on GaAs(110) [7.12], and (3) naphthalene on Pt(111) [7.13–15]. Strongly chemisorbed molecules with rings in their structure appear to be particularly well suited for high resolution STM imaging, although the orientation and bonding of the molecule to the substrate can be a determining factor in the feasibility of obtaining high resolution images [7.11].

7.2.1 Coadsorbed Benzene and CO on Rh(111)

Low energy electron diffraction (LEED) and high resolution electron energy loss spectroscopy (HREELS) measurements had previously shown that the system of benzene (C_6H_6) and CO coadsorbed on Rh(111) is extremely stable, with the molecules strongly chemisorbed in a saturated layer [7.16]. Because the strong chemisorption limited the diffusion rate of the molecules on the surface, STM imaging of the molecules was possible using a room temperature ultrahigh vacuum (UHV) STM which has in situ sample and tip transfer to a UHV surface analysis system [7.17]. Samples were prepared by exposing the cleaned Rh(111) crystal to a fixed dosage of CO gas, with the amount depending on the overlayer structure desired, and then saturating the surface with benzene from the vapor phase [7.8, 9]. Observed ring-like features were associated with individual benzene molecules for molecules arranged in two different ordered overlayers, the (3×3) [7.8] and the $c(2\sqrt{3} \times 4)$rect structures [7.9, 10]. For the latter structure, CO molecules were spatially resolved for the first time. These results first demonstrated the ability of the STM to distinguish between two types of chemisorbed molecules, an important step towards real space imaging of chemical reactions at surfaces. Images also showed translational and rotational domain boundaries, molecules adsorbed near step edges, and evidence for surface diffusion.

The structure of the (3×3) ordered superlattice of coadsorbed benzene and CO on Rh(111), as determined by dynamical LEED analysis [7.18], is shown in Fig. 7.1. The unit cell contains one benzene molecule lying flat and two upright CO molecules, all chemisorbed over hcp-type threefold hollow sites directly over second layer Rh atoms. Figure 7.2 shows threefold symmetric ring-like structures which were associated with individual benzene molecules, although the CO molecules were not clearly resolved [7.8]. The observed threefold symmetry of the molecular images is presumably due to the interaction of the π-bonds of the benzene molecules with the rhodium substrate atoms below; the lobes of the observed benzene features are situated over the bridge sites between two rhodium atoms. The highest resolution images were obtained for electrons

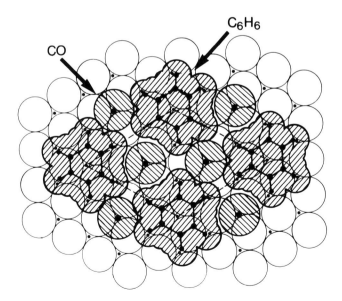

Fig. 7.1. The structure of Rh(111)–$(3 \times 3)(C_6H_6 + 2CO)$ determined from dynamical LEED analysis [7.18]. Large circles and small dots represent the first and second layer metal atoms respectively. From [7.8]

Fig. 7.2. STM constant current topographic image of benzene coadsorbed with CO in a (3 × 3) array on Rh(111), with $V_T = -0.010\,\text{V}$ and $i_T = 2\,\text{nA}$. The threefold ringlike benzene structures are approximately 2 Å high. From [7.8]

tunneling from the tip to the sample at very low voltages. Thus, electrons were tunneling into mostly metallic empty states near the Fermi level, which are expected to have only small contributions from hybridization with molecular states [7.19]. Other images of this structure showed translational domain boundaries and molecules near step edges [7.8]. Images measured about ten minutes apart also displayed evidence of benzene molecular diffusion, with molecules tending to shift into their favored (3 × 3) lattice positions.

The primitive unit cell of the $c(2\sqrt{3} \times 4)$rect overlayer, shown in Fig. 7.3, was determined by dynamical LEED analysis to contain one flat-lying benzene molecule and only one CO molecule, each chemisorbed over an hcp-type threefold hollow site [7.20]. Figure 7.4 shows three-dimensional views of two images of this overlayer with rotational domain boundaries and atomic steps [7.10]. Figure 7.4a shows all three possible rotational domains in the same image. The step edge acts as one domain boundary (R, S), while the other domain boundary (R) intersects the step edge. Step edges often separate such domains, as seen in Fig. 7.4b. Translational domain boundaries and isolated defects have also been observed.

Figure 7.5 shows two views of the highest resolution image of the $c(2\sqrt{3} \times 4)$rect overlayer, with clearly resolved CO and benzene molecules [7.9]. In Fig. 7.5a, the gray scale top view of the data has been overlaid by a mesh with large (small) diamonds representing the top (second) layer Rh atoms. When the benzene rings, which are easily identified from their spacing, were placed on hcp-type three-fold hollow sites according to the LEED model [7.20], the small protrusions in the image occurred exactly at the positions expected for CO in the unit cell. The three-dimensional view in Fig. 7.5b shows more clearly the relative

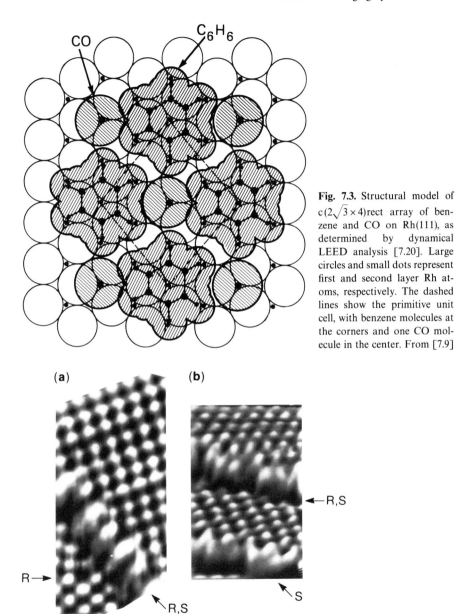

Fig. 7.3. Structural model of $c(2\sqrt{3}\times 4)$rect array of benzene and CO on Rh(111), as determined by dynamical LEED analysis [7.20]. Large circles and small dots represent first and second layer Rh atoms, respectively. The dashed lines show the primitive unit cell, with benzene molecules at the corners and one CO molecule in the center. From [7.9]

Fig. 7.4. (a) Three-dimensional STM image showing three rotational domains in the $c(2\sqrt{3}\times 4)$rect benzene overlayer, with rotational domain boundaries (R) near an atomic step (S) of the Rh substrate. The rotational boundary at the left is also indicated by a line. The $\sim 50\times 85\,\text{Å}^2$ image was measured with $V_T = -0.91\,\text{V}$ and $i_T = 2\,\text{nA}$, and the benzene molecules appear to be $\sim 0.40\,\text{Å}$ high. (b) Three-dimensional view of STM image of benzene molecules showing rotational domain boundary (R) at step edge (S). The $\sim 60\times 45\,\text{Å}^2$ image was measured with $V_T = -0.02\,\text{V}$ and $i_T = 3\,\text{nA}$. From [7.10]

(a)

(b) CO C₆H₆

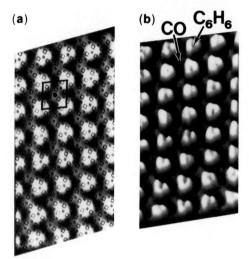

Fig. 7.5. (a) Top view gray scale image of benzene coadsorbed with CO in a c$(2\sqrt{3} \times 4)$rect array on Rh(111), with $V_T = -0.010$ V and $i_T = 2$ nA. The mesh, with large (small) diamonds indicating top (second) layer Rh atoms, has been overlaid on the data according to the dynamical LEED model [7.20]. The primitive unit cell is shown by solid lines, with benzene molecules at the corners and one CO molecule at the filled circle. (b) Three-dimensional view of same data showing clearly both threefold symmetric benzene molecules and small CO protrusions. From [7.9]

height of the benzene molecules, ~ 0.6 Å, compared to the 0.2 Å high CO protrusions. The benzene molecules appear to be threefold symmetric, as they did for the (3×3) overlayer, presumably due to the chemical binding of the molecules to the rhodium substrate. As features associated with both benzene and CO molecules are observed, the STM must be sensitive to localized states of these molecules which are hybridized with the metallic states to yield states near E_F.

7.2.2 Copper-Phthalocyanine on Cu(100) and GaAs(110)

The first STM images which showed the internal structure of isolated molecules were obtained for copper phthalocyanine (Cu-phth) molecules adsorbed in a flat orientation on Cu(100) with two different rotational orientations [7.11]. Cu-phth was sublimed from a quartz crucible onto the cleaned, room temperature Cu(100) crystal. Figure 7.6 shows a model of the Cu-phth molecule over a Cu(100) surface. Tip-induced motion of isolated molecules was occasionally observed. Unusual molecular binding sites at step edges were observed for the first time, which is an important first stage towards studying the chemical reactivity of such sites. Although an isolated molecule was observed on the atomically resolved Cu(100) substrate, giving insight into the molecular binding site, such resolution is difficult to obtain because of strong tip–surface interactions accompanying small tunneling gaps. At coverages close to one mono-

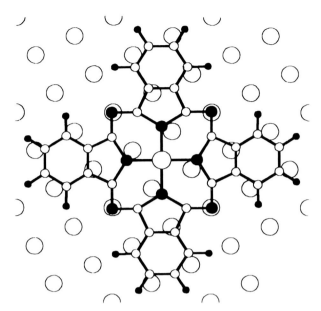

Fig. 7.6. Model of copper-phthalocyanine molecule above a Cu(100) surface. Small (large) circles are C(Cu) atoms and small (large) filled circles are H(N) atoms. The Cu(100) lattice is shown rotated by 26.5°. From [7.11]

layer (ml), the two domains evident in LEED [7.21] are both observed, with each of them corresponding to one of the molecular rotational orientations which are present at low coverages.

Figure 7.7a, measured at low bias, shows the Cu-phth molecules as delocalized pedestals with weak internal structure, with the fine structure emphasized by baseline subtraction. Simple Hückel molecular orbital calculations of the charge distributions of the highest occupied (HOMO) and lowest unoccupied molecular orbitals (LUMO) of the free molecule were performed. A grayscale representation of the HOMO was embedded in the image in Fig. 7.7a. The inequivalent appearance of the two rotational orientations of the molecules suggests a strong tip asymmetry. Figure 7.7b shows an image of a stepped region, with gray levels chosen to highlight the fine structure within the individual molecules. Here the two rotational orientations appear to be symmetric, and the fine structure appears as four protrusions on each of the four lobes of the molecule. This image agrees surprisingly well with the calculations, in which the four predominant protrusions per lobe arise from highly populated carbon atom p-states.

The choice of substrate critically influences the ability of the STM to obtain high resolution STM images of individual molecules. Registry of the molecules with specific symmetrical binding sites on the substrate appears to be helpful in obtaining reproducible, recognizable molecular images. Good STM images of Cu-phth were not obtained for coverages higher than 1 ml on Cu(100), or for submonolayer coverages on Au(111) or Si(111). Instabilities in high coverage images may have been associated with molecular adsorption on the tip, resulting

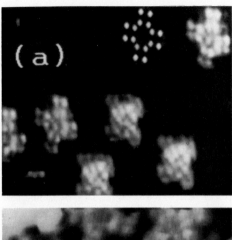

Fig. 7.7. (a) High resolution image of Cu-phthalocyanine molecules on Cu(100) at submonolayer coverage for $V_T = -0.15\,\text{V}$ and $i_T = 2\,\text{nA}$. Fine structure has been emphasized by baseline subtraction, and a gray scale representation of the HOMO, evaluated 2 Å above the molecular plane, has been embedded in the image. (b) High resolution image near 1 ml coverage for $V_T = -0.07\,\text{V}$ and $i_T = 6\,\text{nA}$, with gray levels chosen to highlight fine structure within individual molecules. From [7.11]

in poorly conducting layers which reduce the tip–molecule gap. For Au(111), low activation barriers allowing rotations or translations of molecules may contribute to smearing of the images.

High resolution imaging in UHV of Cu-phth, deposited on GaAs(110), has also been obtained by *Möller* et al. [7.12]. In this study, the molecules were deposited by transferring them with the tunneling tip from one sample, a polycrystalline silver surface onto which 0.5 ml of Cu-phth had been sublimed, to a second sample of freshly cleaved GaAs(110). On GaAs(110), individual molecules display four maxima in the topographic image, which were associated by the authors with the hydrogen atoms in the benzene rings.

7.2.3 Naphthalene on Pt(111)

Recently, high resolution images of naphthalene ($C_{10}H_8$) chemisorbed on Pt(111) have been observed at room temperature in UHV [7.13–15]. LEED measurements on this system had previously shown disordered overlayers forming at room temperature and ordered layers occurring for saturated coverage when the substrate was maintained at temperatures between 100 and 200° C during deposition [7.22]. The ordered overlayer has a (6×3) LEED pattern with certain missing beams attributed to glide plane symmetries. The proposed real space model had a unit cell consisting of two planar molecules, each located at (3×3) Pt lattice points, with molecules in alternating rows having different rotational orientations in a herringbone arrangement.

For the STM measurements, samples were prepared by exposing the cleaned Pt(111) crystal to naphthalene vapor from a gas handling line maintained at 75°C. The substrate was either at room temperature or cooling down from a high temperature anneal to produce disordered and ordered overlayers respectively, as judged by LEED. The room temperature STM measurements were all performed in the constant current topographic mode, with bias voltages of about $+ 1V$ applied to the sample and tunneling current of approximately 1 nA [7.13–15].

High resolution STM images, such as Fig. 7.8a, show naphthalene molecules appearing as bi-lobed structures, with three rotational orientations on the surface, 120° apart [7.13]. For the ordered overlayer, molecules are situated on (3×3) Pt lattice sites in small domains, typically < 50 Å diameter, with domain boundaries involving a molecular shift by an additional Pt lattice constant, as indicated in the schematic diagram in Fig. 7.8b. Although very few of the proposed herringbone (6×3) unit cells are observed, the overlayer arrangement possesses a high degree of order, with $\sim 40\%$ of the molecules in the observed (3×3) domains satisfying the required glide plane symmetry operations and $\sim 30\%$ of the molecules displaying correlated orientations at 6 times the Pt

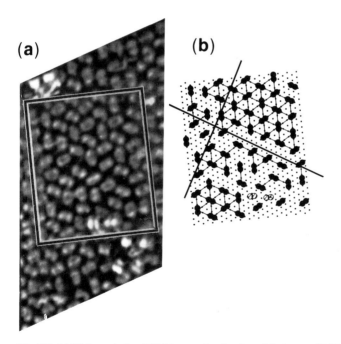

Fig. 7.8. (a) High resolution STM image of ordered naphthalene on Pt(111), $\sim 90 \times 150$ Å² in size, acquired with $V_T = - 0.822$ V and $i_T = 2.5$ nA. The image skewing is due to the correction for severe thermal drift at the time of the data collection. (b) Schematic diagram of the overlay of a Pt(111) lattice with molecular positions. Glide symmetries are indicated by dashed lines. From [7.13]

spacing [7.13]. Thus, the orientational correlations between the molecules in the pseudo-(6 × 3) domains, as seen in the STM images, appear to be high enough to account for the observed LEED pattern.

Bright defects, about twice the typical 1 Å naphthalene height, are apparent in Fig. 7.8a. They are shown as hatched molecules in Fig. 7.8b and appear to have neighboring molecules situated closer together than the usual (3 × 3) lattice spacing; thus, they have tentatively been assigned as naphthalene molecules tilted on edge. Images of the disordered naphthalene overlayer obtained through room temperature adsorption show vastly increased numbers of such high defects [7.13]. Thus, the difference between disordered and ordered overlayers of naphthalene may be associated with thermally enhanced motion of the molecules allowing the system to better accommodate adsorption with the molecular rings parallel to the surface.

STM images of a low coverage (0.2 ml) of naphthalene on Pt(111) indicate that the molecules still have only three rotational orientations, but the molecules are uniformly distributed over the surface with little orientational order, as shown in Fig. 7.9 [7.14]. This image was measured after annealing the sample with adsorbed molecules to less than the 200°C molecular desorption temperature. While no significant aggregation of the molecules occurred, more defects were observed, perhaps as a result of some molecules beginning to decompose. Motion of the molecules between subsequent STM images measured 10 min apart, as in Fig. 7.9a and b, was frequently observed for the low coverage surface but not within domains of the ordered overlayer. The molecules undergo both discrete translational motion (T) from one binding site to an adjacent one and discrete angular changes (R) among the three fixed orientations, probably as a result of a low barrier to such motions at room temperature when there is sufficient space around the molecules.

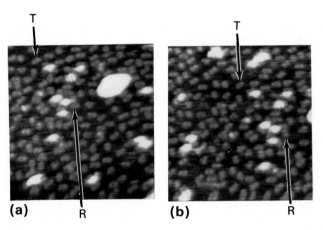

Fig. 7.9. STM image of annealed sample of 0.2 ml naphthalene on Pt(111), with images (**a**) and (**b**) measured 10 min. apart. Note typical rotated (R) and translated (T) molecules. $V_T = -0.51$ V and $i_T = 0.8$ nA. From [7.14]

7.3 STM of Alkanes and Their Derivatives

Many studies of molecules have involved the use of highly oriented pyrolytic graphite (HOPG) as a substrate because it has large, atomically flat terraces, and atomic resolution on the bare graphite surface is readily attainable in an STM operating in air [7.23]. Nonetheless, because graphite is extremely inert chemically, molecules can diffuse easily on the surface, making them difficult to image. Also, the low binding strength of the molecules to the surface makes their interactions with the tunneling tip more likely. Finally, it now appears that artifacts in the STM images of bare graphite can sometimes appear like the images of molecules [7.24].

7.3.1 Cadmium Arachidate and Other Langmuir–Blodgett Films

The first report of STM imaging of cadmium arachidate bilayers on HOPG [7.4], measured in air, involved samples which were deposited by the Langmuir–Blodgett (LB) technique [7.5]. Since graphite is hydrophobic, one layer is deposited on the sample as it is lowered vertically into the water containing the molecules, and a second layer is added as the sample is raised out of the solution (Fig. 7.10). The molecules are assumed to be transferred by this technique so that they are oriented perpendicular to the surface, with the thickness of the bilayer being 54 Å. Bright elliptical spots in the STM constant height images, ~ 3 Å in diameter, were assigned to individual molecules, arranged in an approximately triclinic unit cell, with 5.84 Å along the a-axis and ~ 4 Å along the b-axis, with somewhat irregular molecular spacing (Fig. 7.11). Recently, *Smith* and *Frommer* have ascribed the 3 Å elliptical features in the STM images of cadmium arachidate to the radius of gyration of the hydrocarbon terminus of the lipid molecule [7.25]. Such images on cadmium arachidate and other LB films on graphite have since been reproduced by several groups [7.26–28]. Besides confirming the earlier study, *Hörber* et al. [7.26] showed images of di-myristoyl-phosphatic acid (DMPA) on three different substrates: cleaved graphite, oxidized graphite, and gold-coated calcite in air. By checking the measured lattice spacings for (1) the same LB material and different substrates and (2) the same substrate and different molecules, they ruled out artifacts in the images which might be related to imaging of the graphite substrate with altered and distorted magnification.

Smith et al. [7.4] observed rapid motion of the molecular layer across the viewing field of the STM and speculated that the cadmium arachidate layer might be sliding across the graphite substrate during the imaging. They also reported seeing only graphite substrate images for some areas of the coated sample, indicating nonuniformity of the molecular film. After performing Raman and Fourier transform infrared (FT-IR) spectroscopies on polydiacetylene (PDA), *Rabe* et al. [7.29] concluded that the hydrocarbon chains on graphite, desposited by the LB technique, were mostly oriented parallel to the

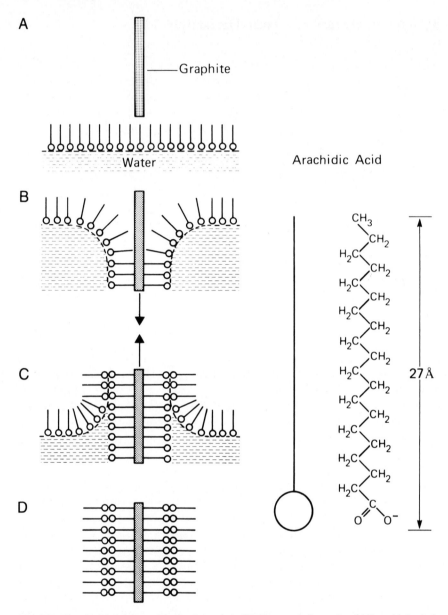

Fig. 7.10 Sample fabrication with the Langmuir–Blodgett technique. Arachidic acid (cadmium arachidate) dissolved in chloroform is spread on a water surface (**A**). The graphite substrate is lowered into the water, and because it is hydrophobic, it gains a monolayer on the way down (**B**). A second monolayer is transferred on the way up (**C**), giving a cadmium arachidate bilayer on graphite in air (**D**). From [7.4]

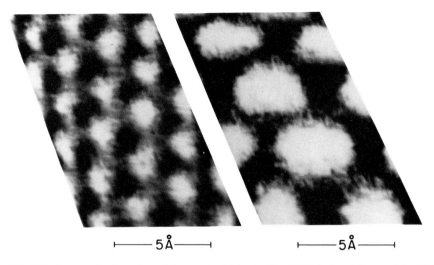

├──────5 Å──────┤   ├──────5 Å──────┤

Fig. 7.11. Images of a cleaved graphite sample obtained with a STM. Left, pristine graphite. Right, graphite coated with cadmium arachidate bilayer. Brightness indicates higher tunneling current, and therefore represents higher topography. From [7.4]

substrate; thus, it is possible that very little of the surface consists of the expected vertically oriented molecular layer, making it more difficult to observe features associated with such an arrangement [7.30]. *Mizutani* et al. [7.27] also commented on the low reproducibility of their STM images of adsorbed cadmium arachidate molecules, as most of their data showed images of the carbon atoms of the bare graphite substrate. They speculate on the possibility that conductive contamination hills may be needed to pin the molecular film on the slippery graphite substrate. Such a hill might also keep the tip farther from the nonconductive molecules and help the imaging of the molecules in the constant height (variable current) mode. Also, the tunneling tip, when it falls on regions of the bare substrate, may brush away the molecules from some areas.

Although several authors [7.4, 26–28, 31] have speculated on possible conductivity mechanisms which would allow the STM imaging of the 54 Å thick cadmium arachidate bilayer, the puzzle still remains. One report even shows images of structures on solid layers of alkanes of various chain lengths, up to a thickness of 1000 Å, on a variety of substrates [7.31]. The typical STM current of 1 nA corresponds to $\sim 10^{10}$ electrons per second, which is 10–14 orders of magnitude higher than would be expected from the measured electrical conductivity of planar metal/fatty-acid-monolayer/metal junctions [7.32]. One possible mechanism is the injection of electrons into loosely bound non-localized orbitals of the normally insulating molecule, which might occur if the work function of the tip were locally much reduced by a foreign adsorbate [7.26]. Another model, involving electron conductance by hopping, requires the existence of a conductive level in the layer, such as may be provided by conductive dusts seen in some

STM images [7.27]. Another possible explanation is resonant tunneling from the Fermi level of the tip through surface states of the LB film which are located near the middle of the electron energy gap [7.31].

7.3.2 n-Alkanes on Graphite

Images of n-alkanes physisorbed onto HOPG have been obtained by *McGonigal* et al. [7.33]. Images were found both for $n\text{-}C_{32}H_{66}$, deposited sequentially from solutions of iso-octane and n-decane, and $n\text{-}C_{17}H_{36}$, deposited from a drop of pure material. The filtered image of $n\text{-}C_{32}H_{66}$ is shown in

1 nm ⊢──┤

Fig. 7.12. Filtered STM image of $n\text{-}C_{32}H_{66}$ molecule adsorbed on graphite. The field of view is approximately $100 \times 100\,\text{Å}^2$, with $V_T = -0.4\,\text{V}$ and $i_T = 1\,\text{nA}$. From [7.33]

(a)

(b)

Fig. 7.13a, b. Schematic diagram depicting $n\text{-}C_{32}H_{66}$ molecules on graphite lattice. (a) Small dots correspond to the graphite sites predominantly imaged by the STM with no adsorbed layer present. The $n\text{-}C_{32}H_{66}$ molecules have been superimposed with their proposed commensurate registration. (b) Large dots mark the graphite sites enhanced by the presence of the $n\text{-}C_{32}H_{66}$ molecules, thus corresponding to the bright area in the experimental image in Fig. 7.12. From [7.33]

Fig. 7.12. The molecules adsorb flat on the surface in an ordered two-dimensional array which is commensurate with the graphite lattice. The adsorbed molecules appear to locally enhance the tunneling current, with dark troughs between the rows being areas without molecular coverage. The spacing and angles of the observed atomic features, however, are those of the substrate graphite atoms, rather than being associated directly with structures in the adsorbed molecules. On bare graphite, every second atom in the lattice is usually imaged by the STM, so that the observed atomic images show features 2.5 Å apart in a hexagonal array [7.34]. Thus, the molecules influence the electronic structure of the combined molecule-substrate system in such a way that the tunneling current is enhanced from certain graphite sites over which the molecules are adsorbed (Fig. 7.13). Unlike the image of n-$C_{32}H_{66}$, the image of n-$C_{17}H_{36}$ shows all of the graphite tunneling sites which would be expected from a clean substrate, although the presence of the molecules again enhances the tunneling from sites over which the molecules are located.

7.3.3 Alkylbenzenes on Graphite

A molecular monolayer adsorbed at the interface between HOPG and an organic solution of didodecylbenzene [$H_{25}C_{12}(C_6H_4)C_{12}H_{25}$] (DDB), concentration 10 mg/ml, has also been directly imaged in situ by *Rabe* and *Buchholz* [7.35, 36]. Figure 7.14 shows two molecular domains, containing lamellae with equal width, which are separated by a molecularly sharp boundary. This image was measured at quasi-constant height in the variable current mode with an average current of 2 nA and tip voltage between +1.2 and +1.5 V. The

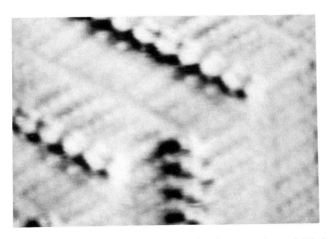

Fig. 7.14. STM image of a domain boundary in a monolayer of didodecylbenzene adsorbed at the interface between an organic solution and the basal plane of graphite. The strongest contrast results from the phenyl stacks. The side chains are tilted relative to the lamella boundary and parallel to each other in the adjacent domains. Image size is 72×47 Å^2. From [7.35]

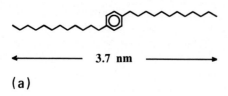

← ———— **3.7 nm** ———— →

(a)

lamella

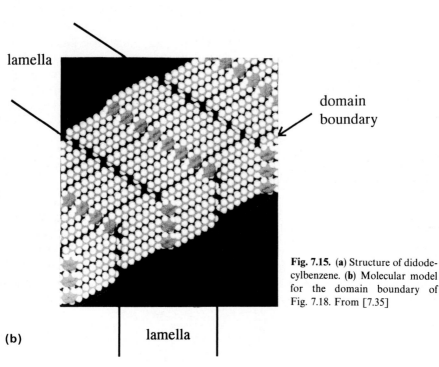

domain
boundary

(b) | lamella |

Fig. 7.15. (a) Structure of didode-cylbenzene. **(b)** Molecular model for the domain boundary of Fig. 7.18. From [7.35]

molecules are extended and lie flat on the HOPG, with higher contrast observed at the positions of the benzene rings, although the alkyl side chains are also resolved. Figure 7.15 shows schematic diagrams of a DDB molecule and the molecular model for the arrangement of molecules in Fig. 7.14. The molecules within a domain pack in lamellae with the long axis of the molecules forming an angle of 60° with the lamella boundary. The adjacent domains consist of molecules in identical conformations, but differing by a 180° rotation of the molecule about its long axis, leading to a symmetrically inequivalent orientation of the molecule relative to the substrate. Domain boundaries such as those in Fig. 7.14 are quite common in this system, with few lamellae larger than about thirty molecules. The motion of such domain boundaries, involving cooperative motion of some molecules near the boundary, has been directly imaged in STM images with fast scans (1 ms per line), recorded on videotape in real time [7.35].

7.4 STM of Liquid Crystals

Liquid crystals are rod-like molecules with phases which are more highly ordered than a liquid but are less ordered than a crystal [7.37]. Nematic liquid crystals show one-dimensional ordering with alignment along a common axis, while smectic liquid crystals order in layers, with each layer having two-dimensional ordering, and no registry between molecules of adjacent layers. Such molecules appear to self-organize on inert substrates such as graphite and MoS$_2$, forming stable, highly ordered two-dimensional molecular structures for STM imaging.

The first successful STM images of liquid crystal molecular systems were presented by *Foster* and *Frommer* [7.38]. They imaged both 4-*n*-octyl-4'-cyanobiphenyl (8CB) and 4-(*trans*-4-*n*-pentylcyclohexyl)benzonitrile (5CBN) on HOPG; at room temperature, the former is a smectic liquid crystal in the bulk while the latter is a supercooled nematic liquid crystal. After placing a drop of the liquid crystal on graphite and waiting several hours, a procedure which may have given the liquid crystal time to self-order on the graphite surface, images of the liquid crystals were obtained in the constant current mode for tip voltage bias of -0.5 to -0.7 V and tunneling current of 0.2 to 0.5 nA. These images showed two-dimensional ordering of the liquid crystal molecules on the graphite substrate, with resolution high enough to resolve individual molecular axes spaced ~ 5 Å apart.

7.4.1 Alkylcyanobiphenyls

Since *Foster* and *Frommer's* first images of liquid crystals were published [7.38], numerous liquid crystals have been imaged with the STM. The 4-*n*-alkyl-4'-cyanobiphenyls (*m*CB, where *m* = 6, 8, 10, or 12) have been particularly well studied by several different groups [7.39–46]. The structure of the 8CB molecule is given in Fig. 7.16.

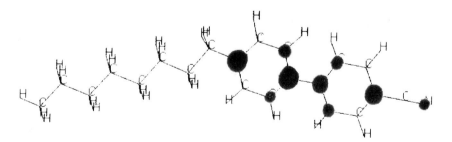

Fig. 7.16. Schematic diagram showing structure of 8 CB molecule. Superimposed is the calculated HOMO for the free 8 CB molecule, showing the surface with molecular orbital density $\varphi^2 = 8.1 \times 10^{-3}$ (Bohr radius)$^{-3}$. From [7.44]

Three different methods have been used for depositing these liquid crystals onto graphite and similar van der Waals surfaces, such as MoS_2, and all yield similar STM images [7.46]. (1) A droplet of liquid crystalline material is placed onto the freshly cleaved surface and then heated into the isotropic phase so that it spreads over the surface and wets it. (2) The molecules at $\sim 100°C$ are sublimed from the vapor phase onto the surface. (3) The molecules are dissolved in a polar solvent, such as alcohol, and the dilute solution is placed onto the substrate, where the solvent evaporates.

These molecules all form ordered structures on graphite, arranging themselves in orderly rows [7.39, 40, 42–44, 46]. Figure 7.17 shows a particularly high resolution STM image and associated model for 8CB on graphite [7.44]. The bright regions in the image appear to be biphenyl rings, and the alkyl tails appear to be somewhat darker. The cyano groups point inward towards each other, and the alkyl tails radiate outward in either a clockwise or counterclockwise direction. The cyano groups interdigitate slightly to increase the packing density and may be very slightly tilted. For 8CB and 12CB, the molecules form unit cells of 8 molecules, while unit cells are composed of 10 molecules for 10 CB.

The particular form of the ordered molecular overlayer also depends on the substrate. When 8CB is adsorbed on MoS_2, a "single row" structure, with cyanobiphenyl head groups and alkyl tails alternating in each row, is formed [7.25, 41, 45]. The unit cell of this structure consists of only 4 molecules and is quite different from that for the same molecule adsorbed on graphite. The alignment of the molecules relative to the underlying substrate lattice was determined by piercing the molecular layer with the tip and imaging the substrate for tunneling gap resistance $< 10^7 \Omega$ [7.44, 45]. For both graphite and MoS_2 substrates, the alkyl chains align with the substrate lattice, similar to the ordering of the n-alkanes described above [7.33]. This alignment appears to determine the form of the ordered molecular overlayer. For the molecule adsorbed on graphite, the alkyl chain of each molecule aligns with a graphite lattice vector, while the cyano groups align along a different vector, as shown in the model in Fig. 7.17b.

The aromatic groups consistently appear to be much higher than the alkyl tails in the STM images, with the height variations being too large to be completely topographic in origin. Calculations of the electronic structure of the free 8CB molecule have been performed using ab initio Hartree–Fock molecular orbital calculations [7.44]. A similar approach had yielded good agreement with STM data for Cu-phth on Cu(100) [7.11]. For the 8CB, the aromatic carbons and the nitrogen had the highest electronic densities for the LUMO and HOMO (Fig. 7.16). These calculations suggest that the origin of the contrast between the aromatic rings and the alkyl tails is related directly to their differing molecular orbital densities near the Fermi level. This explanation conflicts with the argument presented by *Spong* et al. [7.7], which suggested that the contrast in liquid crystal STM images originated from modulation of the local work function of the substrate by the polarizable molecular adsorbates.

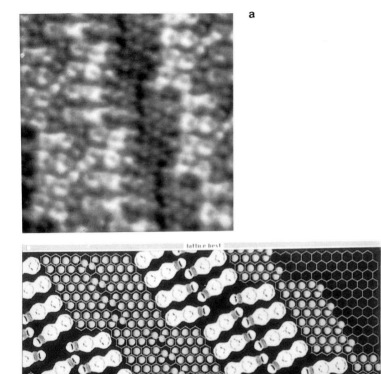

a

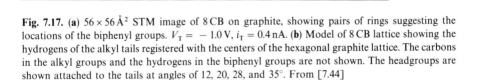

b

Fig. 7.17. (a) $56 \times 56 \, \text{Å}^2$ STM image of 8 CB on graphite, showing pairs of rings suggesting the locations of the biphenyl groups. $V_T = -1.0 \, \text{V}$, $i_T = 0.4 \, \text{nA}$. (b) Model of 8 CB lattice showing the hydrogens of the alkyl tails registered with the centers of the hexagonal graphite lattice. The carbons in the alkyl groups and the hydrogens in the biphenyl groups are not shown. The headgroups are shown attached to the tails at angles of 12, 20, 28, and 35°. From [7.44]

7.4.2 Other Liquid Crystals

A number of other liquid crystals on HOPG have also been successfully imaged by STM. These include 5-nonyl-2-n-nonoxylphenylpyrimidine (PYP909) [7.7, 39], 5-nonyl-2-n-hexoxylphenylpyrimidine (PYP906) [7.47], 4-cyano-4'-n-

butoxybiphenyl (M12 or 4OCB) [7.7], 4-cyano-4'-*n*-heptylbiphenyl (K21) [7.43], 4-cyano-4'-*n*-octylbiphenyl (K24) [7.43], and 4-(4'-*n*-propylcyclohexyl)-cyanocyclohexane (CCH₃) [7.7]. In each case the molecules were deposited on the substrate in the form of a droplet of liquid, and some of the samples were subsequently heated to the temperature at which the bulk liquid crystal forms a smectic phase. All of these molecules contain alkyl groups, which may be important in controlling their binding to the graphite surface, as discussed above for the alkylcyanobiphenyls. [7.25]. Again, the alkyl groups always appeared darker in the images than the aromatic, cycloalkane, and cyano groups.

The STM measurements of these molecules were done with the tip submerged in the droplet of non-conducting liquid; tunneling is assumed to occur through a barrier which is only one monolayer thick. The STM successfully resolves individual molecules and often can also resolve different functional groups within the molecules. STM images of all of these molecules displayed two dimensional ordered arrangements on the graphite surface, with molecules in one row in registry with adjacent molecules in the next row. The rows of molecules often are aligned with a principal axis of the underlying graphite lattice. Substantial substrate–adsorbate interactions result in the formation of highly ordered molecular phases at the graphite surface, which are qualitatively different from the bulk phases of the materials in terms of molecular packing, absence of phase transitions, and elevated isotropic transition temperature [7.47].

7.5 STM of Polymers

The STM has recently been used to observe several polymer systems, including poly(octadecyl acrylate) (PODA), poly(methyl methacrylate) (PMMA) [7.48, 49], poly(2-methyl 1-pentene sulphone) (PMPS) [7.49], and polyethylene (PE) [7.50, 51]. Most of these polymers were observed on HOPG, with similar structures also reported for PODA, PMMA, and PMPS on MoS₂ and Au(111) on mica [7.49]. General morphological features, such as fibrils of polymer and lamella structures were easily observed, with images comparable in lateral resolution to those obtainable by transmission electron microscopy (TEM), but with the higher vertical resolution of the STM.

7.5.1 PODA, PMMA, and PMPS on Graphite

Langmuir–Blodgett (LB) films of PODA, PMMA, and PMPS were prepared by dipping HOPG substrates in both horizontal and vertical geometries [7.48]. The structures of these molecules are shown in Fig. 7.18. Submonolayer coverage of PODA on the surface of horizontally dipped graphite samples yielded images both of isolated fibrils and of bundles of fibrils. Figure 7.19a shows an

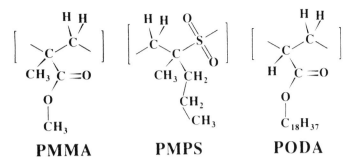

Fig. 7.18. Structures of PMMA, PMPS, and PODA monomer units. From [7.49]

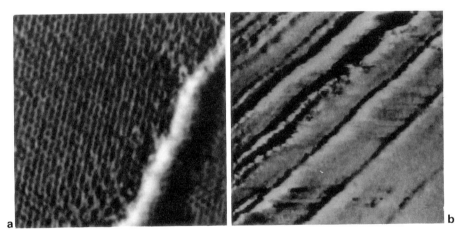

Fig. 7.19a, b. STM images of PODA submonolayers on graphite. (**a**) Isolated fibril on otherwise clean graphite. Image area is $75 \times 75\,\text{Å}^2$. (**b**) In other areas, the much more common fibrils are seen, which are interpreted as bundles of chains. Image area is $440 \times 440\,\text{Å}^2$. From [7.48]

isolated fibril, $8 \pm 2\,\text{Å}$ wide, on the graphite lattice background. It is not clear whether this image shows a single polymer chain or a small bundle of a few chains. The shadow to the right of the fibril is caused by the delayed response of the STM feedback loop to the high current signal obtained over the fibril. Figure 7.19b shows a parallel array of fibrils, which are interpreted as bundles of chains. Very similar images were also observed by atomic force microscopy (AFM) [7.48, 49]. A image of a highly ordered region of PODA, with spacing suggestive of bulk ordering, was also observed [7.48]. Images of syndiotactic PMMA, prepared by vertical dipping, showed several hundred Å domains of parallel polymer fibrils.

Modification of the polymer fibrils could be induced by applying a voltage pulse to the STM tip [7.48, 49]. A rectangular voltage pulse of 100-ns duration broke the fibrils at the place where it was applied and disrupted the parallel fibril structure over a region of radius 50 Å. The voltage threshold for this effect was measured as 4.1 ± 0.2 V and was not polarity dependent, with the peak tunneling current during the pulse of ~ 200 nA. The authors speculate that this effect may be similar to what occurs when PMMA is bombarded with electrons during e-beam lithography, although the energy density is higher than that used for conventional exposure of PMMA and is more comparable with that used for STM-induced lithography on 200 Å thick PMMA films [7.52]. Similar modification of polymer fibrils was also observed for PMPS on graphite [7.49].

7.5.2 Polyethylene on Graphite

The STM has recently been used to examine single crystal polyethylene (PE) lamellae on HOPG [7.50]. These samples were coated with thin Au or Cr overlayers in order to obtain sufficient electrical conductivity for the STM measurements. The Cr layers were thinner (~ 20 Å) and had smaller grain size (~ 5 Å) than the Au layers (~ 100 Å thick, ~ 30–50 Å grain size). Preferential decoration of small step edges, which occurred both at the edges of some PE lamellae and on the freshly cleaved HOPG, were occasionally observed. Isolated individual crystals of PE were also difficult to locate because the PE tended to agglomerate into large mounds on the surface. The features observed on the PE lamellae were mostly on the scale of 100 Å in height and 3 Å laterally, with the morphology of the coated samples varying from single crystal lamellae to large pyramidal structures formed by a multilayer growth process. Such morphological features are well known for such samples and had been previously observed in TEM pictures.

Other recent studies have examined PE crystallized from an entangled solution onto mica substrates and subsequently imaged both by STM [7.51] and by atomic force microscopy (AFM) [7.53]. The STM measurements were performed in air on samples shadowed with a platinum–carbon (Pt/C) mixture and showed some features as small as 7.5 Å, which were not readily interpreted. Both the STM and AFM images showed morphological features similar to those seen in TEM micrographs. Both types of images also showed higher vertical resolution than the TEM micrographs, although they generally had comparable lateral resolution.

7.6 Other Molecules

The surface of a conducting molecular crystal of TTF-TCNQ (tetrathiafulvalene-tetracyanoquinodimethane) was imaged by *Sleator* and *Tycko* [7.54] at ambient temperature and pressure. The observed corrugations and step heights were consistent with the bulk crystal structure of TTF-TCNQ. Individual organic molecules at the surface were successfully resolved. Good correspondence was found for simulated STM images obtained by calculating surfaces of constant probability density using the HOMO of TTF and the LUMO of TCNQ.

The adsorption and ordering of the simple polar molecules acetone and dimethylsulphoxide (DMSO) on graphite have also been observed with the STM [7.55]. After prolonged exposure of graphite to acetone vapor, a 2×2 ordered monolayer forms on the surface. An ordered, incommensurate layer of DMSO, with periodicity approximately 1.4 times that of the graphite, was formed after 35 min. The adsorption and desorption of both types of molecules was followed in the STM images as a function of temperature.

7.7 Conclusions

The images shown here demonstrate that the STM is beginning to be an excellent tool for understanding the surface structure of adsorbed molecules on an atomic scale. The strengths and weaknesses of the technique are still being explored. A large number of scientists are now working in the field, and developments are occurring rapidly in applying the technique to all sorts of molecular systems. (See Chap. 3 of Vol. II for a discussion of biological molecules.) In the future, the applications of the STM to metal–adsorbate surfaces may permit the observations of surface chemical processes, such as molecular diffusion, nucleation phenomena, and step- or defect-related reactivity. Images of other ordered monolayers, such as alkanes, liquid crystals, and polymer materials, are just beginning to yield useful results on the structure of the films and their binding to the substrate.

Acknowledgments. I would like to thank V.M. Hallmark, J.P. Rabe, and D.P.E. Smith, for helpful discussion.

References

7.1   R.J. Behm, W. Hosler, E. Ritter, G. Binnig: Phys. Rev. Lett. **56,** 228 (1986)
7.2   E. Ritter, R.J. Behm, G. Potsche, J. Wintterlin: Surf. Sci. **181,** 403 (1987)
7.3   J.K. Gimzewski, E. Stoll, R.R. Schlittler: Surf. Sci. **181,** 267 (1987)
7.4   D.P.E. Smith, A. Bryant, C.F. Quate, J.P. Rabe, Ch. Gerber, J.D. Swalen: Proc. Natl. Acad. Sci. USA **84,** 969 (1987)
7.5   K.B. Blodgett: J. Am. Chem. Soc. **57,** 1007 (1935)
7.6   D.P.E. Smith, M.D. Kirk, C.F. Quate: J. Chem. Phys. **86,** 6034 (1987)
7.7   J.K. Spong, H.A. Mizes, L.J. LaComb Jr., M.M. Dovek, J.E. Frommer, J.S. Foster: Nature **338,** 137 (1989)
7.8   H. Ohtani, R.J. Wilson, S. Chiang, C.M. Mate: Phys. Rev. Lett. **60,** 2398 (1988)
7.9   S. Chiang, R.J. Wilson, C.M. Mate, H. Ohtani: J. Microsc. **152,** 567 (1988)
7.10 S. Chiang, R.J. Wilson, C.M. Mate, H. Ohtani: Vacuum **41,** 118 (1990)

7.11 P.H. Lippel, R.J. Wilson, M.D. Miller, Ch. Wöll, S. Chiang: Phys. Rev. Lett. **62**, 171 (1989)

7.12 R. Möller, R. Coenen, A. Esslinger, B. Koslowski: J. Vac. Sci. Technol. A **8**, 659 (1990)

7.13 V.M. Hallmark, S, Chiang, J.K. Brown, Ch. Wöll: Phys. Rev. Lett. **66**, 48 (1991)

7.14 V.M. Hallmark, S. Chiang, Ch. Wöll: J. Vac. Sci. Technol. B **9**, 1111 (1991)

7.15 S. Chiang, D.D. Chambliss, V.M. Hallmark, R.J. Wilson, Ch. Wöll: The *Structure of Surfaces III*, ed. by S.Y. Tong, M.A. Van Hove, X. Xide, K. Takayanagi, Springer Ser. Surf. Sci. Vol. 24 (Springer, Berlin, Heidelberg 1991)

7.16 C.M. Mate, G.A. Somorjai: Surf. Sci. **160**, 542 (1985) and references therein

7.17 S. Chiang, R.J. Wilson, Ch. Gerber, Y.M. Hallmark: J. Vac. Sci. Technol. A **6**, 386, (1988)

7.18 R.F. Lin, G.S. Blackmann, M.A. Van Hove, G.A. Somorjai: Acta Crystallogr. **B43**, 368 (1987)

7.19 E.L. Garfunkel, C. Minot, A. Gavezzotti, M. Simonetta: Surf. Sci. **167**, 177 (1986); E.L. Garfunkel: Private communication

7.20 M.A. Van Hove, R.F. Lin, G.A. Somorjai: J. Am. Chem. Soc. **108**, 2532 (1986)

7.21 J.C. Buchholz, G.A. Somorjai: J. Chem. Phys. **66**, 573 (1977)

7.22 D. Dahlgren, J.C. Hemminger: Surf. Sci. **109**, L513 (1981); **114**, 459 (1982)

7.23 G. Binnig, H. Fuchs, Ch. Gerber, H. Rohrer, E. Tosatti: Europhys. Lett. **1**, 31 (1986)

7.24 C.R. Clemmer, T.P. Beebe, Jr.: Science, **251**, 640 (1991)

7.25 D.P.E. Smith, J.E. Frommer: Ordered organic monolayers studied by tunneling microscopy, *Scanning Tunneling Microscopy and Atomic Force Microscopy in Biology*: ed. by O. Marti (Academic, New York 1991)

7.26 J.K.H. Hörber, C.A. Lang, T.W. Hänsch, W.M. Heckl, H. Möhwald: Chem. Phys. Lett. **145**, 151 (1988)

7.27 W. Mizutani, M. Shigeno, K. Saito, K. Watanabe, S. Sugi, M. Ono, K. Kajimura: Japan. J. Appl. Phys. **27**, 1803 (1988)

7.28 H. Fuchs: Phys. Scripta **38**, 264 (1988)

7.29 J.P. Rabe, M. Sano, D. Batchelder, A.A. Kalatchev: J. Microsc. **152**, 573 (1988)

7.30 J.P. Rabe: Private communication

7.31 B. Michel, G. Travaglini, H. Rohrer, C. Joachim, M. Amrein: Z. Phys. **B76**, 99 (1989)

7.32 E.E. Polymeropoulus, J. Sagiv: J. Chem. Phys. **69**, 1836 (1978) and references therein

7.33 G.C. McGonigal, R.H. Bernhardt, D.J. Thomson: Appl. Phys. Lett. **57**, 28 (1990)

7.34 I.P. Batra, N. Garcia, H. Rohrer, H. Salemink, E. Stoll, S. Ciraci: Surf. Sci. **181**, 126 (1987)

7.35 J.P. Rabe, S. Buchholz: Phys. Rev. Lett. **66**, 2096 (1991)

7.36 S. Buchholz, J.P. Rabe: J. Vac. Sci. Technol. **B9**, 1126 (1991)

7.37 P.G. deGennes: Liquid crystals, in *Concise Encyclopedia of Solid State Physics*, ed. by R.G. Lerner, G.L. Trigg (Addison-Wesley, Reading, MA) pp. 143–148

7.38 J.S. Foster, J.E. Frommer: Nature **333**, 542 (1988)

7.39 J.S. Foster, J.E. Frommer, J.K. Spong: Proc. SPIE **1080**, 200 (1989)

7.40 D.P.E. Smith, H. Hörber, Ch. Gerber, G. Binnig: Science **245**, 43 (1989)

7.41 M. Hara, Y. Iwakabe, K. Tochigi, H. Sasabe, A.F. Garito, A. Yamada: Nature **344**, 228 (1990)

7.42 T.J. McMaster, H. Carr, M.J. Miles, P. Cairns, V.J. Morris: J. Vac. Sci. Technol. A **8**, 672 (1990)

7.43 W. Mizutani, M. Shigeno, Y. Sakakibara, K. Kajimura, M. Ono, S. Tanishima, K. Ohno, N. Toshima: J. Vac. Sci. Technol. A **8**, 675 (1990)

7.44 D.P.E. Smith, J.K.H. Hörber, G. Binnig, H. Nejoh: Nature **344**, 641 (1990)

7.45 D.P.E. Smith, W.M. Heckl: Nature **346**, 616 (1990)

7.46 D.P.E. Smith: J. Vac. Sci. Technol. **B9**, 1119 (1991)

7.47 J.K. Spong, L.J. LaComb Jr., M.M. Dovek, J.E. Frommer, J.S. Foster: J. de Phys. **50**, 2139 (1989)

7.48 T.R. Albrecht, M.M. Dovek, C.A. Lang, P. Grütter, C.F. Quate, S.W.J. Kuan, C.W. Frank, R.F.W. Pease: J. Appl. Phys. **64**, 1178 (1988)

7.49 M.M. Dovek, T.R. Albrecht, S.W.J. Kuan, C.A. Lang, R. Emch, P. Grütter, C.W. Frank, R.F.W. Pease, C.F. Quate: J. Microsc. **152**, 229 (1988)

7.50 R. Piner, R. Reifenberger, D.C. Martin, E.L. Thomas, R.P. Apkarian: J. Polymer Sci. C: Poly. Lett. **28**, 399 (1990)

7.51 D.H. Reneker, J. Schneir, B. Howell, H. Harary: Poly. Commun. **31,** 167 (1990)
7.52 M.A. McCord, R.F.W. Pease: J. Vac. Sci. Technol. **B4,** 86 (1986)
7.53 R. Patil, S.-J. Kim, E. Smith, D.H. Reneker, A.L. Weisenhorn: Poly. Commun. **31,** 455 (1990)
7.54 T. Sleator, R. Tycko: Phys. Rev. Lett. **60,** 1418 (1988)
7.55 J.S. Hubacek, R.T. Brockenbrough, G. Gammie, S.L. Skala, J.W. Lyding, J.L. Latten, J.R. Shapley: *J.* Microsc. **152,** 221 (1988)

8. STM on Superconductors

P.J.M. van Bentum and *H. van Kempen*

With 21 Figures

Soon after the tunneling technique was introduced by *Giaever* in 1960 [8.1, 2], it became clear that this would become the method of choice to study the quasiparticle density of states in superconductors. Not only could the energy gap be measured with great accuracy, but detailed predictions from the microscopic Bardeen–Cooper–Schrieffer (BCS) theory of superconductivity [8.3], such as the temperature dependence of the gap and depairing effects in magnetic fields, could also be verified. Tunneling studies on materials with a strong electron–phonon coupling, such as Pb, revealed direct information about the Eliashberg function $\alpha^2(\omega) F(\omega)$, and subsequent modifications of the BCS theory could explain the enhanced energy gap in these strong coupling superconductors. With the prediction and observation of the possibility of pair tunneling [8.4] between two superconductors, a new field of low-temperature physics and practical applications was opened. Since then the field of superconductive tunneling spectroscopy has evolved to include many features, such as proximity effects, non-equilibrium states induced by large injection currents, inelastic tunneling spectroscopy etc. Because of the very strong non-linearity of the tunneling I–V characteristics, associated with the singularity in the density of states just above the gap, superconductive tunnel junctions have been used as microwave radiation detectors or mixers working with nearly 100% quantum efficiency. An overview of the various theoretical aspects and experimental applications of planar tunnel junctions can be found in [8.5–9].

With the advent of the STM many groups recognized the potential of having a tunneling probe without the need to deposit an insulating barrier. It allows the study of "difficult" materials such as organic superconductors or high-T_c superconductors. An important advantage of the STM is that spectroscopic information about the superconductor can be obtained with a lateral resolution of a few ångstroms. Especially for inhomogeneous states in granular superconductors or type II superconductors in a magnetic field, one is able to map the superconducting properties on a length scale smaller than the superconductive coherence length (typically between 10 and 1000 Å).

Nevertheless, it appears that progress in this field is relatively slow because of the technical constraints of a cryogenic environment, in which vibration isolation and in situ surface preparation and characterization is more difficult. A second problem is that the most interesting materials, such as the recently discovered high-T_c superconductors [8.10–12], often have ill-defined surfaces that can even be insulating or semiconducting due to oxygen diffusion. Classical superconductors such as Al, Pb or In also oxidize rapidly in air, preventing high-resolution imaging. A notable exception is the layered superconductor $NbSe_2$, which can easily be cleaved and which is relatively inert for contaminations. Low-temperature STMs have only recently been introduced into a UHV environment with the appropriate manipulation tools or in situ growth techniques to provide a clean and well-defined surface of the superconductor.

In the following we will briefly summarize the theoretical models that are appropriate for tunneling into superconductors. We will concentrate on spectroscopy of the quasiparticle density of states. Since most of the experimental work is not yet at a level of refinement that allows phonon spectroscopy or inelastic tunneling spectroscopy to be done on a routine basis, we refer to the extensive literature on classical planar junction tunneling for these phenomena.

The experimental work on classical superconductors will be described, with emphasis on the spatially resolved determination of the quasiparticle density of states both in the bulk and near a vortex core in a type II superconductor.

A separate section is devoted to spectroscopy on high-T_c superconductors. Since the discovery of the new classes of oxidic superconductors, the interaction causing superconductivity in these materials has been a mystery. Initially it was not even clear if they could be described within the framework of the classical pairing theory. Tunneling spectroscopy can give detailed information about many basic questions and it is obvious why a vast majority of the experimental STM work has concentrated on these materials. Nevertheless, there is still no general consensus about even the most fundamental questions, such as the exact value of the gap in the various materials.

On the other hand, a more comprehensive picture is emerging from the data and most of the physical phenomena can at least be identified and compared with model calculations. We will summarize the central experimental results on topography and spectroscopy and discuss the various interpretations.

8.1 Theory of Tunneling into Superconductors

After the first tunneling experiments on superconductors, *Giaever* [8.1, 2] suggested that the I–V characteristics directly reflected the quasiparticle density of states in the superconductor. In a phenomenological "golden rule" approach, one may expect that the transition probability from the normal metal into the superconductor is proportional to the number of filled states in the normal metal times the number of empty electronic states in the superconductor at the same energy. The net current (corrected for the reverse process) can then be written as

$$I_{NS}(V) = G_{NN} \int_{-\infty}^{+\infty} \varrho_N(E)\varrho_S(E + eV)[f(E) - f(E + eV)]dE , \qquad (8.1)$$

where $\varrho_S/\varrho_N = |E|/\sqrt{(E^2 - \Delta^2)}$ is the normalized BCS density of states, ϱ_N is the electronic density of states of the normal metal at the Fermi level, G_{NN} $= R_{NN}^{-1}$ is the normal state conductance and $f(E)$ is the Fermi–Dirac function. In this semi-phenomenological derivation Giaever assumed that the tunneling probability is independent of energy, in other words that the applied bias voltage is small compared to the Fermi energy, and that the single particle tunneling matrix elements do not change whether the metal is superconducting or not. A more elaborate justification of this approach was obtained using a transfer Hamiltonian approach by *Bardeen* [8.13] and by *Cohen* et al. [8.14]:

$$H_T = \sum_{k,q} (T_{kq} c_k^\dagger c_q + T_{qk} c_k c_q^\dagger) . \tag{8.2}$$

In the formalism of second quantization the electron operators are to be replaced by the Bogoliubov–Valatin operators [8.15, 16] for the quasiparticle excitations in the superconductor:

$$H_T = \sum_{k,q} [T_{kq} (u_k \gamma_{ek0}^\dagger + v_k \gamma_{hk1})(u_q \gamma_{eq0} + v_q \gamma_{hq1}^\dagger)$$

$$+ T_{qk}^* (u_q \gamma_{eq0}^\dagger + v_q \gamma_{hq1})(u_k \gamma_{ek0} + v_k \gamma_{hk1}^\dagger)] . \tag{8.3}$$

The current through the barrier can be calculated from the rate of change of the expectation value of the number of electrons on either side evaluating the commutator relation

$$J_{op} = e\dot{N}_L = \frac{e}{i\hbar}[N_L, H_T] , \tag{8.4}$$

where $N_L = \sum_k c_k^\dagger c_k$. The evaluation of this commutator is rather cumbersome due to four-electron operators that arise from the interaction Hamiltonian in the superconductor. Fortunately, however, a Hartree–Fock approximation can be used and the expression for quasiparticle tunneling reduces to

$$J_{op} = e\frac{2\pi}{\hbar} \sum_{k,q} |T_{qk}|^2 \{u_q^2 [f(\varepsilon_k) - g(E_q)]\delta(E_q - \varepsilon_k)$$

$$+ v_q^2 [f(\varepsilon_k) - (1 - g(E_q))]\delta(E_q + \varepsilon_k)\} , \tag{8.5}$$

where g is the modified Fermi–Dirac distribution function: $g(E_q) = f(|\varepsilon_q|)$, E_q is the excitation energy with respect to the Fermi energy in the superconductor, ε_k is the excitation energy in the normal metal, and $v_k^2 = 1 - u_k^2$ is the BCS coherence factor which gives the probability to find the pair state $(k, -k)$ occupied. In this summation contributions for $q < k_F$ and $q' > k_F$ can be taken together such that $E_q = E_{q'}$ and the BCS coherence factors will vanish because $u_q^2 + u_{q'}^2 = 1$ and $v_q^2 + v_{q'}^2 = 1$. Rewriting the summation in k space as an energy integral will bring back the phenomenological equation (8.1). Following the same procedure for tunneling between two superconductors, *Josephson* [8.4] found that there are in principle two contributions to the tunnel current that are proportional with the square of the transfer matrix element $|T|^2$, one being the quasiparticle term discussed above, and the second a pair tunneling term which is related to the phase difference between the two superconducting condensates. The amazing fact arises that a pair process occurs with a $|T|^2$ probability instead of the expected $|T|^4$. This is due to the fact that the initial and final states are coherent, in contrast to the incoherent nature of single-particle tunneling into excited quasiparticle states.

Figure 8.1 illustrates the two tunneling channels in a normal-metal–superconductor (NS) junction into a $q < k_F$ and a $q' > k_F$ state with the same excitation energy. Figure 8.2 illustrates the single particle tunneling channels between two superconductors. Channel A depicts the tunneling of an existing quasiparticle at $T > 0$ and channel B shows a similar process where a Cooper pair is broken up on the left and a new pair on the right is formed by recombination of the tunneling particle with a pre-existing quasiparticle. In channel C a pair on the left is broken and two quasiparticles are formed on either side of the barrier (possible only for $eV > \varDelta_1 + \varDelta_2$ at $T = 0$). The reverse of process C will occur for negative bias voltage. Finally, in Fig. 8.3 the Josephson process for pair tunneling is illustrated.

In an STM geometry the phase coherence that is necessary for Josephson tunneling is usually destroyed by external perturbations. According to *Anderson* [8.17] the coupling energy between two superconductors is proportional to $\varDelta E = -\hbar J_1(0)/2e$, where the maximum dc supercurrent $J_1(0)$ is typically of order $\varDelta/eR_N$. The tunneling area of an STM junction is extremely small, leading

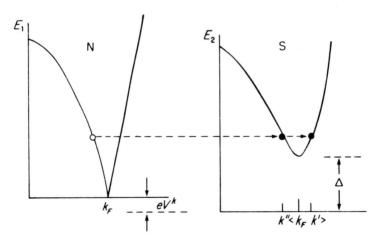

Fig. 8.1 Schematic representation of a tunneling process from an electronic state on the normal metal side ($k < k_F$) into a quasiparticle state with either $k'' < k_F$ or $k' > k_F$ on the superconductor side

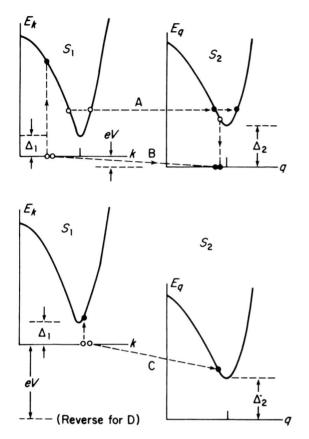

Fig. 8.2. Schematic representation of the various single particle tunneling processes between two superconductors. Process A gives the various tunneling events from an excited quasiparticle state on the left into an empty quasiparticle state on the right. In process B a Cooper pair is broken up on the left, thereby creating a quasiparticle excitation. The second charge carrier recombines with a pre-existing quasiparticle to form a Cooper pair on the right. In channel C a Cooper pair is annihilated on the left and two quasiparticle states are created on both sides of the barrier (only possible for $eV > \Delta_1 + \Delta_2$). The fourth process (D) is the reverse of C

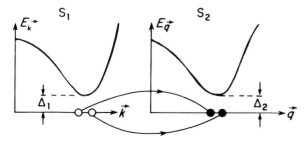

Fig. 8.3. Schematic representation of a Josephson pair tunneling process between two superconductors

to a very small coupling energy which is easily overcome by potential fluctuations that are further facilitated by the very small junction capacitance.

On the other hand, as the barrier thickness in an STM junction can be much smaller than in the traditional case of planar film structures, it is possible that higher order tunneling processes proportional to $|T|^4$ become feasible [8.18]. This possibility was first discussed for the case of planar junctions by *Schrieffer* and *Wilkins* [8.19] to explain the onset of tunneling at $eV = \Delta_1$ or $eV = \Delta_2$ in planar superconductor–insulator–superconductor (SIS) junctions.

A detailed treatment for NS junctions was developed by *Blonder* et al. [8.20].

A fundamental feature in the case of large transmission probabilities is that a pair tunneling process becomes possible. This is known as *Andreev* reflection [8.21]. In this process the initial state can be characterized by an electron excitation on the normal side, while the final state consists of an added Cooper pair in the superconducting condensate and a hole excitation in the normal metal. Energy conservation requires that the initial electron and final hole have symmetric energies with respect to the Fermi level. This process is possible at all bias voltages, giving rise to a sub-gap conductance, and will be strongest for $|eV| \leq \Delta$. In their model *Blonder* et al. constructed a continuous transition from metallic transport to tunneling by introducing a scattering potential of varying height at the NS interface. An interesting aspect of their approach is that they circumvented the problem of an ad hoc assumption as in the tunneling Hamiltonian approach, by matching the solutions of the incoming, reflected and transmitted waves to the Bogoliubov equations on either side of the barrier. In this case the transport of the charges to and from the interface is taken care of in a natural way.

In the case of an NS junction, the result of Blonder et al. can be written in a simple form:

$$J_{NS}(V) = 2N(0)ev_F \int_{-\infty}^{\infty} [f(E - eV) - f(E)][1 + A(E) - B(E)]dE ,   (8.6)$$

where the total transmission probability is determined by the Andreev reflection probability $A(E)$ and by the probability for normal reflection at the boundary $B(E)$. For energies below the gap and at $T = 0$ one can neglect the direct quasiparticle transmission and $B(E) = 1 - A(E)$. Blonder et al. calculate the Andreev reflection probability for a one-dimensional model, in which the barrier at the interface is replaced by a simple delta function $H\delta(x)$. For convenience they introduce a dimensionless barrier strength $Z = H/\hbar v_F$. For $|E| < \Delta$ the Andreev reflection probability is given by:

$$A(E) = \frac{\Delta^2}{E^2 + (\Delta^2 - E^2)(1 + 2Z^2)^2} .   (8.7)$$

For high barriers, $A(E)$ will be proportional to Z^{-4}, reflecting the 2-particle nature of the process. If Z is low (metallic contact), then $A(E)$ will be of order one for $|E| \leq \Delta$, and the conductivity can be a factor of two higher than in the

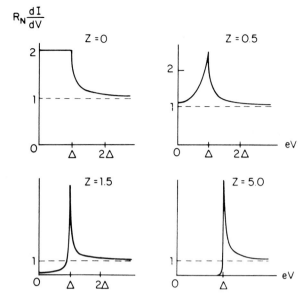

Fig. **8.4.** Tunneling conductance of a low-ohmic junction for different values of the barrier height (Z), allowing for both quasiparticle tunneling and for Andreev reflection. From [8.20]

normal state. The Andreev reflection processes will rapidly decrease for energies above the gap, and the normal quasiparticle tunneling takes over.

In Fig. 8.4 we have reproduced the predicted normalized conductance curves as a function of bias voltage, for different values of the barrier height. For high values of Z the conductance is again proportional to the quasiparticle density of states in the superconductor as in (8.1).

The simple model of Blonder et al. breaks down if the problem can no longer be treated as one dimensional, or if the detailed shape of the tunneling barrier has to be taken into account. In a recent review *Kirtley* [8.22] has summarized the basic theory to deal with low tunnel barriers and possible effects of an anisotropic energy gap.

Following *Simmons* [8.23], one can use a WKB approximation to determine the transmission probability of an energy barrier of arbitrary shape:

$$D(E_z) = \exp\left(-2\int_{s_1}^{s_1} dz \sqrt{2m_z(U(z) - E_z)/\hbar^2} \right), \tag{8.8}$$

where z is the coordinate perpendicular to the plane of the barrier, m_z is the effective mass along z, $U(z)$ the local barrier height, and $s_{1,2}$ are the edges of the barrier, where E_z equals $U(z)$. In the case of an isotropic density of states the tunnel current density can be written as

$$J(V) = \frac{-4\pi m_\parallel e}{h^3} \int_0^{E_m} dE_z D(E_z) \int_0^\infty dE_\parallel (f(E + eV) - f(E))\varrho_S(E). \tag{8.9}$$

The main effect of having an energy-dependent transmission probability is that the I–V characteristics will become nonlinear at voltages comparable to the barrier height. In addition the current will become slightly asymmetric for positive and negative bias voltages. It is instructive to evaluate the tunneling conductance by taking the derivative of this equation:

$$\frac{dJ}{dV}(V) = \frac{-4\pi m_\parallel e}{h^3}\left[\int\limits_0^{E_m} dE_z \frac{dD(E_z, V)}{dV} \int\limits_0^\infty dE_\parallel (f(E + eV) - f(E))\varrho_S(E) \right.$$

$$\left. + \int\limits_0^{E_m} dE_z D(E_z, V) \int\limits_0^\infty dE_\parallel \frac{df(E + eV)}{dV} \varrho_S(E) \right] . \tag{8.10}$$

At $T = 0$, and for a slowly varying transmission probability (that is for barrier heights that are large compared to Δ), we find the very simple result:

$$\frac{dJ}{dV}(V) = \frac{4\pi m_\parallel e^2}{h^3} \varrho_S(E) \int\limits_0^{E_m} dE_z D(E_z, V) . \tag{8.11}$$

This shows that for intermediate barrier heights we still obtain the superconductive density of states except that it is multiplied with a slowly varying transmission function. For extremely low barriers where the transmission probability tends to unity it is more appropriate to use the formalism of Blonder et al. as discussed above.

8.1.1 Coulomb Blockade

For small tunnel junctions with a very small capacitance between the electrodes the electrostatic charging energy of a single electron ($e^2/2C$, where C is the junction capacitance) can be much larger than other relevant energies such as $k_B T$. In such a case the Coulomb charging term in the Hamiltonian will completely dominate the tunneling process [8.24–26]. As a result of this term tunneling is strongly suppressed (Coulomb blockade) when the energy gained by tunneling (eV) is smaller than the energy required to charge the junction.

A detailed theoretical treatment [8.26] shows that, for ideally current biased junctions, at low voltages a quadratic current–voltage dependence is expected while for large voltages the I–V characteristic approaches an asymptote displaced from the origin by $\Delta V = e/2C$. A semi-classical approximation based on a Monte Carlo simulation of the tunneling events [8.27–29] gives the I–V relation also for the intermediate voltage region and for more general source impedances.

A second result of the Coulomb blockade shows up when two small tunnel junctions are connected in series. A tunnel current then can only flow if sufficient voltage is applied to charge the central electrode with one or more electrons. This charge is not a classical variable, but is quantized in units of e. The maximum charge is limited by the applied voltage across both junctions in series, and the tunnel current will increase stepwise on increasing this voltage every time a net extra electron charge can be accommodated on the central electrode. This characteristic shape of the I–V curve is generally referred to as the Coulomb staircase.

The exact shape of the I–V characteristics depends on the ratio of the two capacitances and on the ratio of the tunneling resistances (as long as they remain large compared with the quantum resistance $\hbar/e^2$). In the regime where the ratio of the tunneling resistance is large one could easily mistake the sharp peaks in the conductance at $eV = (2n + 1)e/2C_2$, where C_2 is the larger of the two capacitances, for superconductive energy gaps. Especially near low bias voltages the two types of gap structures look very much alike.

In the case of shallow surface or impurity states in oxide layers one may expect that the local capacitance is sufficiently decoupled from the surroundings that single particle tunneling events give rise to potential oscillations with an appreciable amplitude of the order of e/C. The main effect will be that the probing energy is not well defined and the spectroscopic features will be smeared out. More importantly, the structures will be shifted to higher bias voltages. A similar distortion of the spectroscopic performance could be the result of resonant tunneling through impurity states in the oxide. In order to be sure about the reliability of superconducting gap measurements it is necessary to have a clean surface and tip. A good check is that the Coulomb structures will depend on the distance between tip and surface, while the superconductive gap structure should remain at the same bias voltage.

8.2 Low Temperature STM Spectroscopy on Classical Superconductors

In one of the first experimental low-temperatures STM investigations *Elrod* et al. [8.30, 31] studied polycrystalline Nb_3Sn films at temperatures down to 6 K. After a tip-cleaning procedure by tunneling at a high voltage they were able to measure local variations of the zero bias conductance in a true vacuum tunneling mode. The possible cause for the observed local variations of the superconductive properties is not clear, but it cannot be excluded that they are related to surface damage introduced by ion milling the samples or by the tip-cleaning procedure.

Kirtley et al. [8.32] were the first to combine a topographic scan with a full spectroscopic I–V measurement at each point for NbN films. Although the lateral resolution was limited, the measured gap was clearly position dependent with a relatively large scattering of gap values. Again, the reason for the gap inhomogeneity was not clear and could not be linked directly to details of the surface topography. When a magnetic field of up to 6 T was applied, regions of reduced gaps were measured but there was no evidence of a vortex lattice.

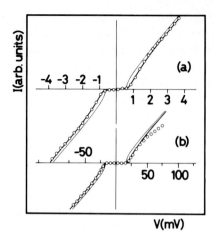

I(arb. units)

-4 -3 -2 -1

(a)

1 2 3 4

(b)

-50

50   100

V(mV)

Fig. 8.5. I–V characteristics (open circles) of point contact junctions on (**a**) Sn and (**b**) YBa$_2$Cu$_3$O$_7$ at 4.2 K. The upper thin line is a BCS prediction using the gap value measured simultaneously in a planar junction geometry. The solid lines are Monte Carlo calculations incorporating a Coulomb blockade

As a test for the capabilities of low-temperature STM spectroscopy on superconductors, *Ramos* et al. [8.33] measured the I–V characteristics by tunneling from a cleaned tungsten tip into a bulk Pb sample both below and above T_c. They found a BCS-like density of states with $\Delta_{Pb} \approx 1.25$ meV at $T = 4.9$ K. At this temperature however, the gap discontinuity is smeared out considerably by thermal broadening.

Wilkins et al. [8.34] used a Au tip to tunnel into a bulk Pb sample. After crashing into the surface, some of the lead stuck to the tip and the presence of superconductor–superconductor (SS) tunneling could be determined from the much sharper discontinuity at eV = 2Δ.

Smokers et al. [8.35] used an evaporated Al–Sn planar tunnel junction to determine the energy gap of the Sn film. Especially when the Al film is also superconducting, this can be done with great accuracy. Simultaneously, they measured the STM I–V characteristics for tunneling between a tungsten tip and the Sn upper film of the planar sandwich. They found that the two gap values were identical, provided that the full I–V characteristic was fitted using the Coulomb blockade model described above. In Fig. 8.5a we have reproduced the STM I–V datapoints (open circles). The thin line is a BCS calculation using $\Delta = 0.59$ meV as measured with the planar junction. The solid line is a Monte Carlo calculation to incorporate the Coulomb blockade, using the same Δ and a junction capacitance $C = 3.2 \times 10^{-16}$ F. Figure 8.5b is a tunneling I–V characteristic on YBa$_2$Cu$_3$O$_7$. A similar fit gives $\Delta = 16$ meV, and $C = 1.6 \times 10^{-17}$ F.

8.3 Vortices

A special case of spatial variation of the superconducting properties occurs when a magnetic field is present perpendicular to the surface of a type-II superconductor. The magnetic field can penetrate the superconductor in quantized magnetic flux lines called vortices. In the vortex core, which contains a single flux quantum, the material is in the normal state. Under proper conditions the vortices are arranged in a regular lattice known as the Abrikosov flux lattice. Vortices have been the subject of many theoretical studies, but until recently, the only observation method was based on decoration techniques [8.36], which confirmed the existence of the Abrikosov lattice but did not give detailed information on the electronic structure. The ability of the STM to perform local spectroscopic measurements makes it an excellent tool for locating the vortices and studying their electronic structure.

An excellent demonstration of the usefulness of STM for research on the flux lattice has been the work of *Hess* et al. on 2H-NbSe$_2$ [8.37–39]. In zero magnetic field they found a quasiparticle density of states in good agreement with BCS.

By measuring the spatial variation dI/dV at a bias voltage where a considerable difference exists between the normal and superconducting state, images of the vortex lattice have been obtained [8.39] at different values of the applied magnetic field (Fig. 8.6). The full potential of an STM local tunneling probe was demonstrated in the measurements of the density of states in and near the vortex core. Instead of a normal, constant conductivity they observed a zero

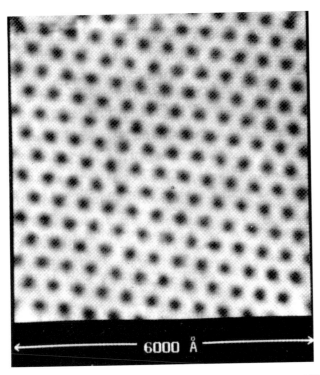

Fig. 8.6. Abrikosov flux lattice in a 1T magnetic field in NbSe$_2$ at 1.8 K. The gray scale corresponds with dI/dV. From [8.39]

bias enhancement. This might be the result of the containment of the quasiparticle wave functions in the potential well around the vortex. Andreev reflection occurs at the positions where the energy of the electron-like quasiparticles matches the local pair potential $\Delta(x)$. This particle in a (cylindrical) box problem will have additional bound energy states. Detailed treatment of the zero bias anomaly in the vortex by *Overhauser* et al. [8.40, 41] and by *Shore* et al. [8.42] predict a double peak structure in the density of states at some distance from the vortex core. This can be interpreted as a result of the higher energy bound states which have a larger spatial extension. In Fig. 8.7 we have reproduced the results of *Hess* et al. [8.37], showing the spatial evolution of the density of states when approaching the vortex core. The zero bias peak at the core position and the development of a double sub-gap structure some distance from the core are clearly seen. For comparison, the zero magnetic field conductance is plotted, showing a well-developed gap with no sub-gap conductance.

In a subsequent investigation, Hess and coworkers found that the tunneling conductance at zero bias decayed most rapidly along the direction connecting two vortices. Along the intermediate directions, rotated by 30° around the vortex center, the decay is much slower. As shown in Fig. 8.8, this leads to a typical star shape for the contours of constant conductance. At a bias voltage of about half the energy gap, the behavior is exactly opposite and the star shape is rotated by 30°. A subsequent analysis by *Gygi* et al. [8.43, 44] showed that

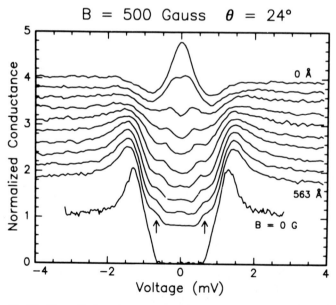

Fig. 8.7. Normalized conductance curves for various distances from a vortex core, showing both the zero bias maximum at the centre of the core and the development of a double structure inside the gap when moving away from the core. The lower trace is a zero field conductance curve. From [8.37]

500 Gauss

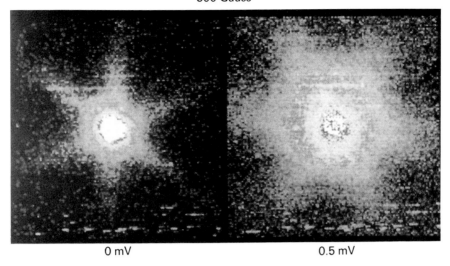

0 mV	0.5 mV

Fig. 8.8. Top view of the conductance dI/dV at $V = 0$ and $V = 0.5\,\text{mV}$ in the plane around a vortex core for an applied magnetic field of 500 Gauss. From [8.37]

breaking the rotational invariance of a vortex, for example by other vortices nearby, leads to an angular band structure for the low-energy part of the quasiparticle spectrum, which is in good agreement with the experimental observations.

The study of vortices is not only important for understanding the fundamental physics of superconductivity but is also relevant for practical applications. For the current-carrying properties of superconductors, the dynamic behaviour of the vortices plays an essential role. An applied magnetic field and current cause a Lorentz force on the vortices. When the force can overcome the pinning forces which can occur at crystal boundaries and imperfections, the vortices start to move. As this vortex movement causes energy dissipation and therefore resistance, one has to understand and control this effect to attain high current densities. This means that methods have to be found to observe the dynamics of the vortices under various conditions.

Dittrich et al. [8.45] demonstrated that it is possible to study the dynamics of vortices with STM. They observed the passage of vortices at a fixed point at the surface of a niobium sheet by placing the tip at that point and measuring the voltage pulses, each of which signals the passage of a vortex. The autocorrelation function of the tip voltage (in non-feedback mode) was used to characterize the average vortex motion. Of course, it is of special interest to perform the same type of measurements on the high-T_c materials. However, the previously mentioned surface preparation problems are a serious handicap. A possible way around this problem would be to use a magnetic force microscope which works at a larger distance from the surface and so is less hindered by contamination layers. One has to be careful however that the magnetic field of the tip does not influence the measurement results. To shed some light on this question *Berthe* et al. [8.46] have compared vortex lattice observations on NbSe$_2$ with tips of W, PtIr and ferromagnetic Ni. They found no detectable influence of the magnetism of the Ni tip on the vortex images. Therefore it can be concluded that magnetic force microscopy can be applied to vortex observations if the flux lattice is sufficiently stiff and/or if the local pinning forces are high enough.

8.4 Organic Superconductors

A final class of superconductors that can be tentatively placed under the heading of conventional superconductors are the organic materials. Atomic resolution topographic images have been obtained by several groups [8.47–50]. *Bando* and coworkers [8.49] measured the I–V and dI/dV characteristics between a Pt:Au tip and (BEDT-TTF)$_2$Cu(NCS)$_2$, which has a T_c of about 11 K. Although the conductance minima at zero bias were only of the order of 50%, there was a clear indication of the opening of a gap with both a BCS-like magnitude and temperature dependence. In Fig. 8.9 we have reproduced the dI/dV curves at

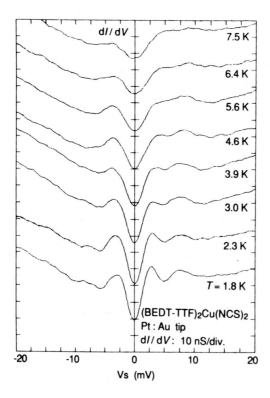

Fig. 8.9. Temperature dependence of the conductance between a Pt:Au tip and the organic superconductor (BEDT-TTF)$_2$Cu(NCS)$_2$. From [8.49]

varying temperatures between 1.8 and 7.5 K. In cases where the surface was not well etched, they observed evidence for a Coulomb blockade with an effective capacitance $C \approx 10^{-17}$ F.

8.5 STM Topography on High-T_c Superconductors

After the spectacular discovery of the new high-T_c superconductors [8.10–12] many SEM and STM investigations were undertaken to study the surface structure. The main aspect of study for the ceramic materials was to examine the influence of the granular structure and twin boundaries on the superconductive properties. Later, as soon as they became available, thin films and single crystals were also subject to intensive study.

8.5.1 Granularity and Growth Structures

Research into the surface structure of granular high-T_c superconductors by STM which is aimed at determining the grain size and the way the grains are connected has been carried out by a number of scientists [e.g. 8.51–54]. A major problem often encountered in working with polycrystalline and even with single-crystalline samples is the presence of an insulating or semiconducting surface layer which prevents tunneling or which necessitates the use of high voltages. The nature of this layer depends strongly on the production method. Another complication is the loss of oxygen under ultra-high vacuum conditions for many of the high-T_c compounds. This is a serious problem especially when the experiments have to be done in oxygen-free conditions, for example at cryogenic temperatures. Several methods have been proposed to get a surface layer which better represents the bulk structure. One method to circumvent the oxygen loss is to cool the samples in an oxygen atmosphere to 100 K before evacuating the sample space [8.51]. One method to get a good surface is to break the superconductor and use the freshly exposed surface (e.g. *Okoniewski* et al. [8.54]). However, this does not always work, because for some of the compounds the surface deteriorates very quickly in air. This deterioration process can sometimes be followed by STM observation [8.55]. Good results are obtained by breaking or cleaving the sample at low temperatures in liquid He [8.56, 57] or at room temperature in a glove box with a very pure atmosphere containing less than 1 ppm of oxygen and water vapour [8.58, 59].

The results on ceramics confirm the granular nature also seen by SEM. Atomic structure is generally not observed with the exception of the broken ceramic $YBa_2Cu_3O_{7-\delta}$ samples of *Okoniewski* et al. and of *Zheng* et al. [8.54, 60] which show periodic features of the size of the unit cell in some flat regions. For thin films a quite granular structure is also often observed [8.61, 62].

Next we discuss observations on thin films and single crystals which show features that might be relevant to the understanding of growth processes. Steps and terraces are often present. The terrace height generally agrees with a multiple of the unit cell dimension (or half a unit cell in e.g. the $Bi_2Sr_2Ca_1Cu_2O_{8-\delta}$ and related compounds which have two equivalent layers in the unit cell) in the c direction, that is the direction which is perpendicular to the largest surface of single crystals [8.52, 63, 64]. The steps are often straight and in that case run parallel to one of the main axes [8.65]. Curved steps are also seen which resemble growth spirals. *Nakamura* et al. [8.66] observed that the direction of the steps of their cleaved film sample is strongly related to the orientation of the (001) MgO substrate. It could be concluded that the films grow such that the a or b axes are aligned with one of the $\langle 100 \rangle$ or $\langle 110 \rangle$ substrate directions. Relating the step height to the unit cell dimension of the different members of the Bi-compound family, *Nakamura* et al. could estimate the relative abundance of the different phases. Both straight and curved steps have also been observed with atomic force microscopy (AFM) [8.65] on $HoBa_2Cu_3O_{7-\delta}$. The advantage is that this method is less critically dependent on the surface quality (Fig. 8.10).

A direct relation between the growth mechanism and external appearance was present in the experiment and analysis of *van de Leemput* et al. [8.67] on $YBa_2Cu_3O_{7-\delta}$. The authors applied a periodic bond chain analysis [8.68] to the crystal structure which enabled them to predict quantitatively the morphology of the single crystals in good agreement with experiment. For example, {013}

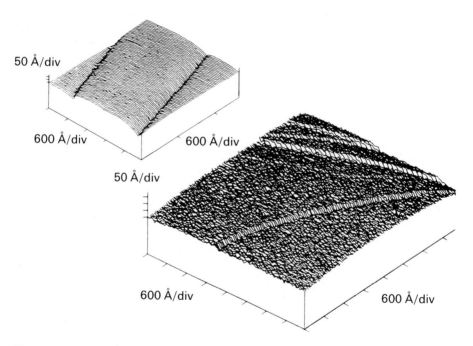

Fig. 8.10. Image of the (001) surface of a $HoBa_2Cu_3O_{7-\delta}$ crystal. The upper figure is taken with an AFM, the lower with STM. From [8.65]

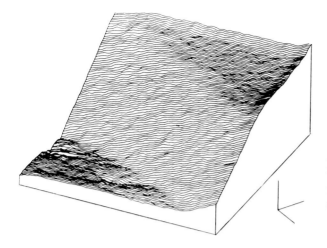

Fig. 8.11. Three facets on an $YBa_2Cu_3O_{7-\delta}$ crystal. The horizontal planes are {001} facets; the sloping plane has been identified as the predicted {013} facet. From [8.67]

facets were observed on the (001) surface in accordance with the theoretical analysis (Fig. 8.11).

Another example of the direct influence of the growth mechanism on the external morphology was observed by *Niedermann* et al. [8.69, 70]. By combining a SEM and STM they could observe in detail the shape of the (100) and (001) faces of the tetragonal precursor phase of $YBa_2Cu_3O_{7-\delta}$ crystals. The different growth mechanisms of the two faces could be identified.

Occasionally local disordered corrugations are observed on the otherwise flat terraces (e.g. [8.65, 71]). These corrugations can be of different size and the origin is unknown. Obviously future research could shed more light on the growth processes.

8.5.2 Potentiometry

For technological applications of high-T_c superconductors the ability to carry a high current density is of crucial importance. This is true not only for applications in power lines etc. but also for thin film devices where, as a result of the small dimensions, the maximum current density can become a limiting factor. It is expected that the current carrying capability is strongly influenced by the granularity of the material. For more fundamental properties such as magnetisation and surface impedance, the granularity can also play an important role. A high-resolution method to study the electrostatic potential along the surface of the current carrying conductors in the normal and superconducting state could be very helpful in elucidating the influence of the granularity.

Muralt and *Pohl* [8.72] have demonstrated that the STM can be used to measure the electrostatic potential along a surface with high spatial resolution. Using a somewhat different method to *Kirtley* et al. [8.73], *Kent* et al. [8.74, 75] measured the potential of current-carrying high-T_c thin films in the normal state to find the correlation between topography and electrical properties. For this purpose a special long-range STM was built. Potential jumps have been observed at well-defined boundaries in the materials, often between clusters of grains with the same potential. A clear difference between post-annealed films and in-situ annealed films could be observed.

The same authors also measured the potential distribution of films in the superconducting state [8.76]. In that case it was necessary to cover the superconducting film with a thin Au film to increase the surface conductivity. It was possible to make this Au film thin enough so that the potential distribution was not disturbed.

8.5.3 Incommensurate Modulation

In this section some atomic or near-atomic resolution results will be discussed. By near-atomic we mean a resolution sufficient to observe rows of atoms but not the individual atoms. Generally the highest resolution is only obtained when the samples are prepared in some protecting atmosphere. A UHV environment can be used to prevent contamination. However, as already mentioned, the oxidic high-T_c compounds have a tendency to lose oxygen in UHV and the surface layer might not be representative for the bulk. This is especially the case for $YBa_2Cu_3O_{7-\delta}$ and related compounds, and for the more stable $Bi_2Sr_2Ca_1Cu_2O_{8-\delta}$ it is suggested that this effect can also play a role [8.63]. Other preparation methods leading to (near) atomic resolution are cleavage in a protecting atmosphere of pure Ar [8.58] or cleavage at liquid helium temperatures [8.56, 57]. Nevertheless near-atomic resolution has sometimes been obtained without these precautions (e.g. $Bi_2Sr_2Ca_1Cu_2O_{8-\delta}$ [8.55], $YBa_2Cu_3O_{7-\delta}$ [8.65]) and even very well-resolved imaging of $Tl_2Ba_2CaCu_2O_8$ appears to be possible in air [8.77].

Amongst the most successful observations are the images made on the $(Bi_{1-x}PB_x)_2Sr_2Ca_1Cu_2O_{8-\delta}$ compounds with x varying from 0 to 0.35. In these compounds a 1D incommensurate superstructure is present, which has also been seen by HREM, X-ray diffraction and electron diffraction (see [8.58] and references therein). Observations have been made on cleaved crystals in a pure Ar atmosphere with $x = 0$ [8.58, 78], $x = 0.13$ and $x = 0.35$ [8.58], on cleaved samples in air with $x = 0$ and $x = 0.16$ [8.79] and on samples cleaved in air and observed under liquid helium with $x = 0$ [8.80]. The results show a periodicity which increases with lead concentration. There is good agreement for $x = 0$ (period = 25–29 Å). The increase with x, however, shows considerable scatter. This is in accordance with results from the other measurement techniques [8.58]. For example, $x = 0.16$ gives 37 and 42 Å [8.79] and $x = 0.15$ gives 32 Å [8.58]. This shows that preparation conditions can apparently influence the period length. The origin of the superstructure is not yet certain. One model assumes an insertion of an extra row of O atoms into the BiO layer every nine to ten Bi sites which causes the BiO layer to buckle. The atomically resolved images of [8.58] are in good agreement with this model. *Kirk* et al. [8.63] conclude from their observations that there are missing rows of Bi atoms while *Shih* et al. [8.81] observe only a displacement of atoms. Some of this confusion might be due to the uncertainty of whether Bi or O atoms are being observed in the BiO layers [8.58]. The possible loss of oxygen in a UHV environment might also play a role [8.58]. For increasing Pb concentration the ordering in the superstructure decreases, and for $x = 0.35$ the superstructure can no longer be seen (Fig. 8.12). The fair agreement of the STM results with HREM, X-ray diffraction and electron diffraction shows that the structure of the surface layer is quite representative of the bulk structure.

The $Tl_2Ba_2CaCu_2O_8$ compound can be imaged with atomic resolution in air [8.77]. For this material a superstructure has also been observed but it is quite weak and without long-range order. Further observation on this material and also on $Tl_2Ba_2Ca_2Cu_3O_{10}$ in pure Ar showed the existence of two phases [8.59], a dominating metallic phase and, for a small fraction of the surface, a semi-conducting phase with a different periodicity.

Fig. 8.12. STM images of (**A**) $Bi_2Sr_2Ca_1Cu_2O_8$, $(Bi_{1-x}Pb_x)_2Sr_2Ca_1Cu_2O_8$ with $x = 0.15$ (**B**) and $x = 0.35$ (**C**). From [8.58]

8.6 STM Spectroscopy on High-T_c Superconductors

As already mentioned at the beginning of this chapter, STM investigations of the newly discovered class of oxidic superconductors [8.10–12] could be of vital importance in determining the nature of the superconductivity and possibly the interaction that causes the high critical temperatures.

In the following we will summarize the various features commonly observed in the tunneling characteristics of both the normal and superconducting state and discuss their possible physical origin.

8.6.1 Normal State Spectroscopy

a) Electronic Structure

An important ingredient for the theoretical modeling of high-T_c materials is the electronic band structure. Yet, despite many experimental and theoretical efforts, much remains uncertain. Even the basic question of whether the conduction can be described in a Fermi liquid picture remains a matter of debate. Various high-energy techniques have revealed the dominant oxygen $2p$ nature of the electronic states at the Fermi level [8.82]. Angle-resolved photo-emission experiments [8.83] indicated the presence of a Fermi edge, although the exact nature of the edge states could not be determined.

Tanaka and coworkers [8.84, 85] and *Shih* et al. [8.81] measured the high voltage tunneling conductance curves on cleaved surfaces of BiCaSrCuO samples and found indications for a small semiconducting gap of the order of 0.3 eV at the surface (Fig. 8.13). Since the surface layer is most probably a BiO layer this would indicate that the metallic conduction, and thus the superconductivity, is mainly located in the CuO_2 layers. *Hasegawa* et al. [8.86] reach a similar conclusion using low-temperature STM spectroscopy, albeit with a somewhat smaller value for the semiconducting gap, which was of the order of 50 meV. The effect of Pb substitution in $(PbBi)_2CaSr_2Cu_2O_8$ was studied by *Wu* and coworkers [8.58, 87]. It appears that the states near the Fermi level are not changed and thus the Pb does not contribute to the electronic states involved in the superconductivity.

The case for $YBa_2Cu_3O_{7-\delta}$ is less clear because the material is more difficult to cleave in situ in order to obtain clean surfaces. Using $YBa_2Cu_3O_7$ crystals that were fractured in an ultra high vacuum, *Chen* and *Tsuei* [8.88] obtained spectra that were indicative of both metallic and semiconducting regions. In this case it was suggested that the BaO layers play the role of the semiconductor. This would be consistent with the general experience of most of the groups that tried to measure the energy gap on the exposed flat *ab* face. In most cases evidence for superconductivity was only obtained after pushing the tip into the surface.

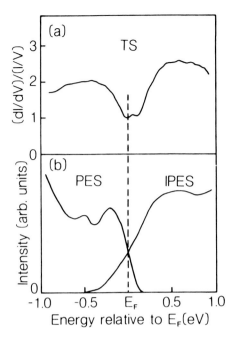

Fig.   8.13a, b. Normalized   conductance $(dI/dV)/(I/V)$ of   a   cleaved   surface   of $Bi_2CaSr_2Cu_2O_8$. For comparison, the results from photoemission (PES) and inverse photo-emission (IPES) are shown. From [8.84]

b) Linear Contribution to dI/dV

A frequently observed feature of tunneling characteristics, even at temperatures above T_c, is the presence of a linear contribution to the conductivity. Also, there is sometimes (but not always) an asymmetrry between positive and negative bias voltages. Several effects have been proposed to explain the linear background conductivity. *Flensberg* et al. [8.89], and *Anderson* and *Zou* [8.90] considered the possibility that this could be proof of a resonating valence bond (RVB) state, where two quasiparticles (a so-called holon and a spinon) must be created when an electron is added to the system. However, as discussed by *Kirtley* [8.22], this explanation is hard to defend because of the large spread of the magnitude of the linear term for different materials and for different junction resistances. In some cases the slope in dI/dV can even become negative [8.91].

Kirtley and coworkers [8.92] initially suggested that the inclusion of small particles in the barrier would explain the V shape of the conductance, as in the model of *Zeller* and *Giaever* [8.24]. However the linear term does not saturate at higher voltages and the model of Zeller and Giaever is not appropriate because it would imply the presence of a very wide distribution of particle sizes, including very small ones. In the more recent treatment of Coulomb blockade effects [8.26] one would expect a linear term for bias voltages up to $e/2C$ and a fast saturation above that value. This behavior was verified in low-temperature STM experiments on Si, where the stray capacitance of the leads is decoupled by the large series resistance in the surface doped layer [8.35]. There are indications that the same occurs in high-resistance junctions on relatively dirty surfaces of high-T_c samples [8.28, 8.93]. However, the Coulomb blockade does not explain the behavior of low-resistance junctions, where the conductivity does not saturate at higher voltages.

To analyse the results on LaSrCuO [8.94], where the linear term is especially prominent, *Kirtley* [8.22] evaluated the classical tunneling expressions described above to include the effect of low tunneling barriers. Among other things, this explains the observed assymmetry in the characteristics. He finds a fair agreement between experiment and theory for barrier heights of the order of 20 meV and for an additional smearing of the gap discontinuity either by lifetime effects or by a Gaussian distribution of local gap values (Fig. 8.14). In YBaCuO the linear term is usually much smaller and a fit to the experimental data is less satisfying [8.22].

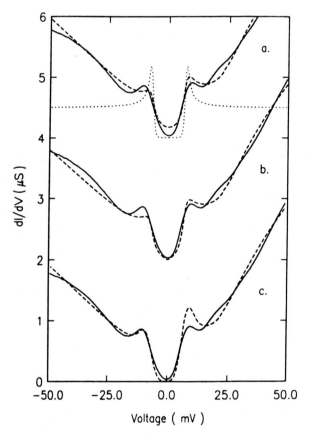

Fig. 8.14. Comparison of point-contact conductance measurements on $LaSr_{2-x}CuO_4$ with various theoretical models. The solid lines are the experimental data from *Kirtley*, the dotted line is the BCS prediction for a gap of 7 meV. The dashed lines are calculations for the case of a very low barrier with different assumptions about the broadening near the gap: (**a**) lifetime broadening, (**b**) strong coupling effects and (**c**) a Gaussian distribution of gaps. From [8.22]

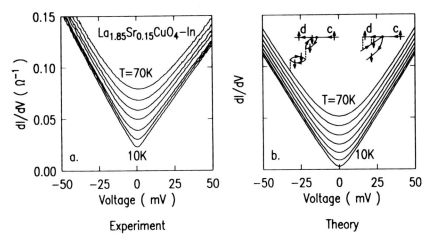

Fig. 8.15. Comparison between the experimental results on $La_{1.85}Sr_{0.15}CuO_4$-In point-contacts by *Kirtley* and *Scalapino* and a theoretical model assuming inelastic scattering due to spin fluctuations in the superconductor. From [8.95]

A third suggestion by *Kirtley* and *Scalapino* [8.95] is the possibility that the linear term reflects the presence of inelastic scattering from a broad continuum of states (Fig. 8.15). From NMR and Raman scattering experiments [8.96–98] there is indeed some evidence for the existence of spin fluctuations in high-T_c superconductors.

Varma et al. [8.100] considered the possible modification of the electronic density of states as being due to interaction between a spin- and a charge-like excitation with a similar energy distribution. Tunneling into such a material could give a strong inelastic contribution in which both types of quasiparticles are excited in the superconductor. Since the final state consists of two quasiparticles, the inelastic tunneling probability will increase with V^2.

At this time it is not possible to decide which of the proposed mechanisms is the dominant one. It may be that several processes coexist, making it hard to draw definite conclusions about the fundamental normal state properties behind the linear term in the conductivity.

8.6.2 STM Spectroscopy of the Superconducting State

The central question in low-temperature tunneling spectroscopy is the magnitude of the energy gap. In the extended BCS scheme, the ratio of this gap to the critical temperature could give some insight into the strength of the electron–phonon coupling. In classical superconductors this ratio $2\Delta/k_B T_c$ ranges between 3.2 and 4.5 from weak- to strong-coupling superconductors respectively. If, however, the gap would be outside this range, this could indicate the presence of an unusual type of superconductivity. Also, if the density of states below the gap is not zero, this could indicate the presence of a pairing state with a different symmetry (p or d wave) in k space. Since these questions are obviously relevant for the theoretical understanding of the high-T_c super-conductivity, many groups initiated low-temperature STM experiments on the new materials. Unfortunately there are many practical problems associated with surface effects on these oxide materials. For the La and Y classes of material the superconducting properties depend strongly on the oxygen content. At the surface and in vacuum there can be an oxygen deficiency at the outer layers of the crystal. In that case, an STM experiment would probe superconducting properties that are different from the bulk. A second problem may arise because of the short coherence length of these superconductors. In classical super-conductors the interaction is averaged over many crystallographic unit cells, and impurities or local defects do not play a significant role. However in the layered materials the coherence length perpendicular to the planes can be less than 1 nm. At the surface, the superconductive order parameter Δ can be considerably suppressed but may sometimes be enhanced. It is not a priori evident that the very local probe technique of STM will give relevant inform-ation about the bulk properties. A third problem is that various growth techniques may leave traces of flux material or other contaminations on the crystal surface. For the La and Y materials it is not easy to use in situ cleavage along the planes. Many of the reported experiments with low-temperature STMs are in fact point-contact experiments where the tip is crashed into the surface. In this case one should not be surprised to find indications of non-local tunneling (either parallel tunneling into non-superconducting surface layers, or series tunneling effects between fractured crystallites). One may also worry about the actual shape of the tunneling barrier. A final point of caution about the interpretation of these point-contact tunneling situations is that the locally exerted pressure can be very large (tens of kilobars) and the superconducting properties may change locally [8.101].

Kirtley [8.22] has given an extensive survey of the tunneling data that are aimed at the determination of the energy gap in high-T_c materials. In the following we will not seek completeness, but rather select some representative examples and discuss the physical interpretations that are given.

8.6.3 Energy Gap

One of the most important problems in STM spectroscopy is to find a consensus between the many seemingly conflicting gap values that are reported. In general the reported results can be divided into four categories.

i) Some of the very early experiments on sintered materials showed ex-tremely high gap values. In some of the more recent tunneling experiments on single crystals or thin films anomalously high gap values are reported with a gap-T_c ratio $2\Delta/k_B T_c = 12$–17 [8.102, 55] or even 37.5 if a reduced surface T_c is taken into account [8.102]. In these cases, it was noted that the zero-bias conductance was lower for characteristics that indicated the highest gap values. This may be a sign that a series of tunnel junctions is formed between several grains in the material, leading to an onset of current at some multiple of the fundamental gap.

ii) The main body of data show gap values between $2\Delta/k_B T_c = 5$–7 with some tendency to decrease to the lower bound as time goes by [8.56, 57, 103–106]. The spreading in results, even on a single sample, can be quite large, the gap discontinuity is usually broadened and the sub-gap conductance can be quite prominent. The same is true for the planar tunneling results [8.107–109] which show weak maxima in the conductance for bias voltages in this range.

iii) A third category of experiments points in the direction of a lower gap, approximately consistent with the BCS value $2\Delta/k_B T_c = 3$–4.5 [8.57, 64, 110, 111]. These values are mostly found for SIS tunneling or in thin film or single crystal investigations where the tunneling direction is along the c axis.

iv) Finally there are some infrequent reports of sharp singularities at a few meV [8.112, 113]. Tentatively these observations might be attributed to the presence of inhomogeneous phases with a lower T_c.

In the following we will restrict the discussion to the second and third categories. The large scattering in results is at least partly due to the different ways in which the spectra are interpreted. Before addressing the problem of the gap value, we will discuss some of the more qualitative features that are prominent in most of the tunneling characteristics.

a) Sub-gap Conductance

In many cases the I–V characteristics do not indicate a real gap, but show a substantial sub-gap conductance. The most far reaching conclusion could be that this indicates the existence of a lower symmetry superconducting state (p or d wave) where there can be nodes in the gap function. At present, however, there seems to be no positive proof for such a hypothesis.

A very elegant proposal is the model of *Tachiki* et al. [8.114]. They start from the assumption that layered materials have essentially a two-dimensional band structure, where the superconductivity is mainly located in the CuO_2 layers. If the transfer interaction between the layers is treated as a weak perturbation, one can calculate the quasiparticle density of states in each layer selfconsistently. The main result of this treatment is that the BaO and CuO layers will have a small proximity-induced energy gap, combined with additional Van Hove singularities at higher energies (Fig. 8.16). At higher temperatures these singularities disappear and a smooth rounded structure remains. In vacuum tunneling on a virgin crystal with an exposed ab plane, the dominant tunneling contribution will come from the BaO or occasionally the CuO planes [8.67], which are weakly metallic, and irregular structures, corresponding to the Van Hove singularities, may result. The energy gap of the CuO_2 layer will only appear near steps or after crashing into the surface. The same will apply for planar junctions

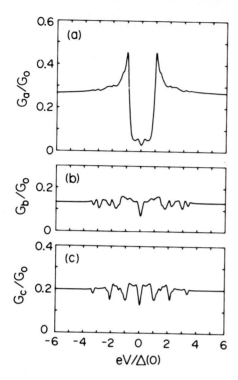

Fig. 8.16. Tunneling conductance curves on $Bi_2Sr_2CaCu_2O_y$ for decreasing tip-sample separation (from **a** to **c**). From [8.86]

where an average is measured over many different surface structures. An interesting experiment in this respect is that of *Hasegawa* et al. [8.86], who measured the tunneling conductance on the *ab* plane of $Bi_2Sr_2CaCu_2O_y$ as the tip approached the surface (Fig. 8.17). At large distances a semiconductor-like gap comparable to that seen at room temperature [8.84] was observed. At very close range a small peak-like structure emerges, which was assigned to tunneling in the lower CuO_2 layer. The peak separation suggests a gap of 28 meV, corresponding to $2\Delta/k_B T_c = 3.6$. After crashing into the surface a zero-bias peak emerges which was attributed to proximity effects. However, this could also be due to Andreev reflection processes, where the zero-bias conductance rises to a maximum of twice the normal state conductance, depending on the interface scattering potential.

A third possibility for sub-gap conductance is proposed by *Phillips* [8.115] who considered the possibility of intrinsic localized states in domains in the CuO_2 planes, that do not contribute to the superconductivity.

Finally, as discussed above it is possible that higher order tunneling processes proportional to $|T|^4$ do play a role in STM junctions on superconductors. Especially in cases where the tip actually touches the surface, one may expect deviations from the usual tunneling characteristics. As calculated in Fig. 8.4 by *Blonder* et al., this will lead to a nonzero conductance below the gap, due to the presence of Andreev reflection processes. A second indication for the presence of

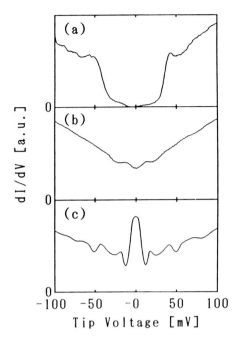

Fig. 8.17. Model calculation of the tunneling conductance for a layered material with proximity coupling between the superconducting (a) and two-dimensional normal metal layers (b) and (c). From [8.114]

higher order tunneling effects can be found in some of the SIS tunneling characteristics [8.93], where there is a small but distinct upturn of the currrent at a bias voltage $eV = \Delta$. This feature is known from the classical literature and was explained by *Schrieffer* and *Wilkins* [8.19] as being due to a $|T|^4$ process. However, in the usual case of relatively high resistance junctions it is expected that the Andreev processes will be very weak, except for bias voltages very close to the gap.

In Fig. 8.18a we have reproduced the I–V characteristic for tunneling between two grains in a thin $YBa_2Cu_3O_{7-\delta}$ film. Figure 8.18b is the BCS prediction for SIS tunneling. Figure 8.18c is an example of a Coulomb staircase measured on a $YBa_2Cu_3O_{7-\delta}$ single crystal and Fig. 8.18d is a theoretical calculation.

b) Gap Broadening

Several attempts have been made to reconcile the broadened singularity in the density of states with a quantitative model. In the model originally constructed by *Dynes* et al. [8.116] a broadening in the density of states is induced by a finite lifetime of the quasiparticles due to inelastic scattering. In a formal way this can be described by introducing a complex quasiparticle energy in the BCS expression for the normalized density of states, $\varrho_S(E) = \mathrm{Re}[|E - i\Gamma|/\sqrt{(E - i\Gamma)^2 - \Delta^2}]$. At present it is not clear what the

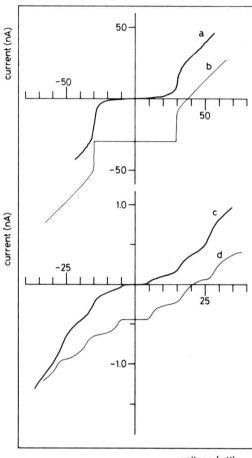

Fig. 8.18. (a) SIS tunneling characteristic between two grains in a $YBa_2Cu_3O_{7-\delta}$ film, (b) BCS prediction for $T = 0$, (c) Incremental single electron charging (Coulomb staircase) of an isolated grain on a $YBa_2Cu_3O_{7-\delta}$ single crystal, (d) Monte Carlo calculation of the Coulomb staircase. From [8.93]

physical nature of this inelastic scattering could be. An alternative approach to describe a life-time broadening of the quasiparticle stress is to use the classical description of strong electron–phonon coupling in which a complex energy-dependent order parameter $\Delta(\omega)$ is introduced. However, one must bear in mind that the broadening is not universal and is for example absent in SIS characteristics on Ti compounds [8.110, 111] (Fig. 8.19).

A second way to interpret the data is to assume a Gaussian distribution of gaps. This could be due to gap inhomogeneities in real space, either perpendicular or in lateral directions. Variations along the perpendicular directions are more likely because the probed surface will in general be smaller than the coherence length in the plane and the superconductive properties will vary most rapidly perpendicular to the layers. In the model of *Tachiki* et al. [8.114] the broadening would be a natural result of the proximity coupling to the nonsuperconducting layers. (Their model is basically a discrete three-layer system. In

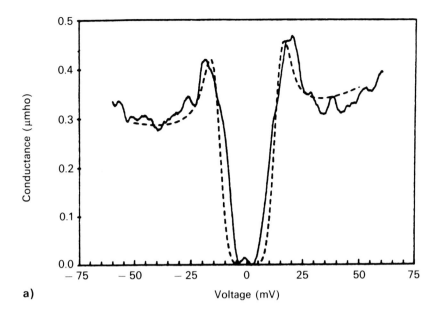

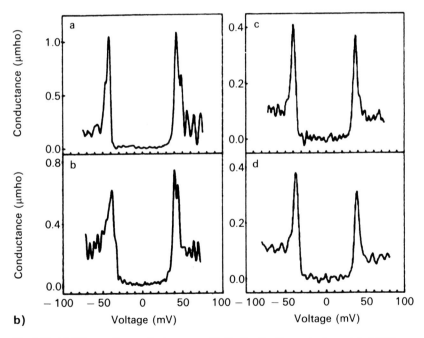

Fig. 8.19. (a) SN tunneling conductance on a $Tl_2Ca_2Ba_2Cu_3O_{10+\delta}$ single crystal at 4.2 K, compared with a calculation assuming $\Delta = 16$ meV, $\delta\Delta = 0.3$ meV and barrier heights $\Phi_1 = 0.5$ eV, $\Phi_2 = 0.2$ eV, (b) SIS tunneling conductance curves observed at different positions on a $Tl_2Ca_2Ba_2Cu_3O_{10+\delta}$ sample, indicating $\Delta \approx 20$ meV. From [8.110]

reality there will be a more continuous variation of the energy gap along the *c* axis).

Kirtley [8.22] proposes an alternative explanation, based on an anisotropic value of the gap in *k* space. Such an anisotropy would not be surprising in view of the very anisotropic crystal structure and also of the transport properties. The broadening in the gap discontinuity in then a natural consequence of the distribution of the *k* vectors of tunneling charge carriers. This proposal is further supported by the difference in energy gaps that are measured when tunneling into the *ab* directions, as compared with those in which the carriers tunnel along the *c* axis.

c) Gap Value and Anisotropy

As summarized above the main body of experiments yield gap values between roughly $2\Delta/k_B T_c = 3$–7. At least part of this scattering is due to the specific analysis of the chosen data. In general it is not sufficient just to take the peak-to-peak distance of the various conductance maxima. Several possible explanations for the observed scattering in gap values have been proposed. Firstly, one cannot exclude completely the possibility of non-local SIS tunneling, which could explain the fact that many experiments show maxima at twice the BCS value. Secondly, the larger gap values on the Bi compounds might be attributed to the presence of a semiconducting gap in the BiO layer. Alternatively, the lower gaps could be due to a proximity-induced gap in the possibly nonsuperconducting BaO or CuO layers, as in the model by *Tachiki* et al. [8.114].

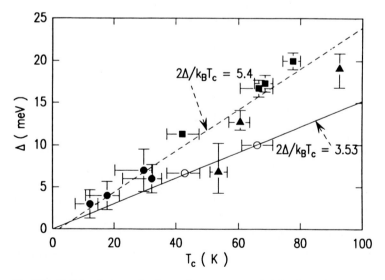

Fig. 8.20. Relation between the observed energy gap and T_c for the various high-T_c materials. For details see text. From [8.22]

Thirdly, as discussed by *Kirtley* [8.22], the scattering of gap data could be related to a fundamental anisotropy of the gap in ***k*** space. In Fig. 8.20 we have reproduced a compilation of gap values from three systematic investigations of the energy gap in the La and Y systems. The solid dots are data from *Fein* et al. [8.117] for $La_{2-x}Sr_xCuO_4$ with varying Sr concentration, the solid triangles are data from *Volodin* and *Khaikin* [8.118] for $YBa_2Cu_3O_{7-\delta}$ and the solid squares are data from break junction tunneling from *Tsai* et al. [8.119] of oriented $YBa_2Cu_3O_{7-\delta}$ films where the carriers tunnel along the *ab* directions. Finally the open circles are data from *Tsai* et al. for tunneling in the *c* direction. We must note, however, that many of these data points are taken from characteristics that show only a weak and relatively broad conduction peak and have a substantial zero bias conductance.

d) Temperature Dependence of the Gap

The evolution of the gap with increasing temperature was measured by *Igushi* et al. [8.120] on planar $YBa_2Cu_3O_{7-\delta}$–Pb junctions of sintered samples. *Lee* et al. [8.121] measured the temperature dependence of $Bi_2Sr_2CaCu_2O_x$–Nb planar junctions. The first temperature-dependent point-contact results were reported

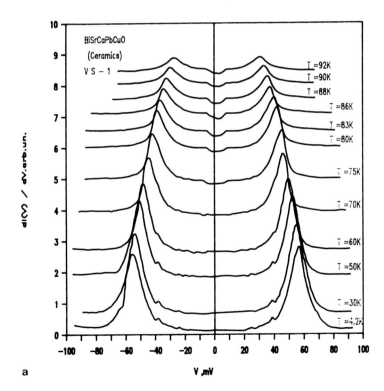

a

Fig. 8.21a. See next page for caption

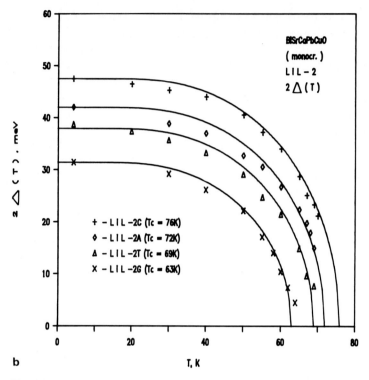

b

Fig. 8.21. (a) dI/dV characteristics for break junctions in BiSrCaPbCuO (2:2:2:3) ceramics at different temperatures ($T_c = 101$ K), (b) Temperature dependence of the gap $2\Delta(T)$ for BiSrCa-PbCuO single crystals with varying T_c. The solid lines correspond to BCS theory. From [8.123]

by *Edgar* et al. [8.122] on sintered $YBa_2Cu_3O_{7-\delta}$–Au junctions. In all cases a good agreement with the predicted BCS temperature dependence was found. More recently, break junction SIS tunneling characteristics were measured as a function of temperature by *Aminov* et al. [8.123], which are reproduced in Fig. 8.21a. The inferred gap values are plotted in Fig. 8.21b as a function of temperature, for different single crystals of $Bi_2Sr_2CaCu_2O_x$ (Pb). The solid lines correspond to the BCS prediction with an enhanced energy gap of the order of $6\,k_B T_c$.

8.7 Concluding Remarks

It is not possible at present to give a final assessment or interpretation of the gap data on high-T_c materials.

Such an assessment may depend on whether reproducible high quality characteristics can be obtained in UHV-compatible STMs with well-characterized clean single crystal samples, if at the same time an atomically resolved surface topography allows one to identify the outer surface layer. Even then, the results could be affected by coherence length effects or oxygen diffusion effects. A meticulous comparison with other techniques that have a larger penetration depth could then be the ultimate test for a comprehensive analysis of the gap.

On the other hand we see an improvement in and a convergence of the experimental results, together with the emergence of quantitative theoretical models that allow a direct comparison with experiment. The stage is set, the players are identified, but maybe the hardest part of the play is yet to begin.

The present picture is that there is a gap in these materials, which may put some of the theoretical models to the test. The c-axis gap of these materials is most probably near the BCS value, while the ab-plane gap is enhanced. The anisotropy is moderate in $YBa_2Cu_3O_{7-\delta}$, but appears to be large in the Bi materials. The anisotropy of the Tl compounds and of NdCeCuO is not yet known. The sub-gap conductance, most prominent in evaporated junctions appears to be most naturally explained in the layer model of *Tachiki* et al. [8.114]. Finally, the temperature dependence seems to be in accordance with the BCS prediction.

References

8.1   I. Giaever: Phys. Rev. Lett. **5**, 147 (1960)
8.2   I. Giaever: Phys. Rev. Lett. **5**, 464 (1960)
8.3   J. Bardeen, L.N. Cooper, J.R. Schrieffer: Phys. Rev. **108**, 1175 (1957)
8.4   B.D. Josephson: Phys. Lett. **1**, 251 (1962)
8.5   L. Solymar: *Superconductive Tunneling and Applications* (Chapman Hall, London 1972)
8.6   W.L. McMillan, J.M. Rowell: Tunneling and strong-coupling superconductivity, in *Superconductivity*, ed. by R.D. Parks (Dekker, New York 1969) p. 561
8.7   C.B. Duke: *Tunneling in Solids*, Solid State Physics, Vol. 10 (Academic, New York 1969)
8.8   E. Burstein, S. Lundqvist (eds.): *Tunneling Phenomena in Solids* (Plenum, New York 1969)
8.9   E.L. Wolf: *Principles of Electron Tunneling Spectroscopy*, International series of monographs on physics **71** (Oxford University Press, New York 1985)
8.10   G. Bednorz, K.A. Müller: Z. Phys. B **64**, 189 (1986)
8.11   M.K. Wu, J.R. Ashburn, C.J. Torng, P.H. Hor, R.L. Meng, L. Gao, Z.J. Huang, Y.Q. Wang, C.W. Chu: Phys. Rev. Lett. **58**, 908 (1987)
8.12   H. Maeda, Y. Tanaka, M. Fukutomi, T. Asano: Jpn. J. Appl. Phys. **27** L209 (1988)
8.13   J. Bardeen: Phys. Rev. Lett. **6**, 57 (1961)
8.14   M.H. Cohen, L.M. Falicov, J.C. Phillips: Phys. Rev. Lett. **8**, 316 (1962).
8.15   N.N. Bogoliubov: Nuovo Cimento **7**, 794 (1958)
8.16   J.G. Valatin: Nuovo Cimento **7**, 843 (1958)
8.17   P.W. Anderson: In *Lectures on the Many-Body Problem*, ed. by E.R. Caianiello (Academic, New York 1964) p. 113
8.18   N. Garcia, F. Flores, F. Guinea: J. Vac. Sci. Technol. A **6**, 323 (1988)
8.19   J.R. Schrieffer, J.W. Wilkins: Phys. Rev. Lett. **10**, 17 (1963)
8.20   G.E. Blonder, M. Tinkham, T.M. Klapwijk: Phys. Rev. B **25**, 4515 (1982)
8.21   A.F. Andreev: Zh. Eksp. Teor. Fiz. **46**, 1823 (1964) [Sov. Phys. JETP **19**, 1228 (1964)]
8.22   J.R. Kirtley: Tunneling measurements of the energy gap in high-T_c superconductors, to appear in Int. J. Mod. Phys. B

8.23   J.G. Simmons: J. Appl. Phys. **34**, 1793 (1963)

8.24   H.R. Zeller, I. Giaever: Phys. Rev. **181**, 789 (1969)

8.25   E. Ben-Jacob, Y. Gefen: Phys. Lett. **108A**, 289 (1985)

8.26   D.V. Averin, K.K. Likharev: J. Low Temp. Phys. **62**, 345 (1986)

8.27   K. Mullen, E. Ben-Jacob, R.C. Jacklevic, Z. Schuss: Phys. Rev. B **37**, 98 (1988)

8.28   P.J.M. van Bentum, H. van Kempen, L.E.C. van de Leemput, P.A.A. Teunissen: Phys. Rev. Lett. **60**, 369 (1988)

8.29   P.J.M. van Bentum, R.T.M. Smokers, H. van Kempen: Phys. Rev. Lett. **60**, 2543 (1988)

8.30   S.A. Elrod, A.L. de Lozanne, C.F. Quate: Appl. Phys. Lett. **45**, 1240 (1984)

8.31   S.A. Elrod, A. Bryant, A.L. de Lozanne, S. Park, D. Smith, C.F. Quate: IBM J. Res. Dev. **30**, 387 (1986)

8.32   J.R. Kirtley, S.I. Raider, R.M. Feenstra, A.P. Fein: Appl. Phys. Lett. **50**, 1607 (1987)

8.33   M.A. Ramos, S. Vieira, A. Buendia, A.M. Baro: J. of Microsc. **152**, 137 (1988)

8.34   R. Wilkins, M. Amman, R.E. Soltis, E. Ben-Jacob, R.C. Jaklevic: Phys. Rev. B **41**, 8904 (1990)

8.35   R.T.M. Smokers, P.J.M. van Bentum, H. van Kempen: Physica B **165-166**, 63 (1990)

8.36   H. Traeuble, U. Essman: J. Appl. Phys. **25**, 273 (1969)

8.37   H.F. Hess, R.B. Robinson, J.V. Waszczak: Phys. Rev. Lett. **64**, 2711 (1990)

8.38   H.F. Hess, R.B. Robinson, R.C. Dynes, J.M. Valles Jr., J.V. Waszczak: J. Vac. Sci. Technol. **A8**, 450 (1990)

8.39   H.F. Hess, R.B. Robinson, R.C. Dynes, J.M. Valles Jr., J.V. Waszczak: Phys. Rev. Lett. **62**, 214 (1989)

8.40   A.W. Overhauser, L.L. Daemen: Phys. Rev. Lett. **62**, 1691 (1989)

8.41   A.W. Overhauser, L.L. Daemen: Phys. Rev. **B40**, 10778 (1989)

8.42   J.D. Shore, M. Huang, A.T. Dorsey, J.P. Sethna: Phys. Rev. Lett. **62**, 3089 (1989)

8.43   F. Gygi, M. Schluter: Phys. Rev. Lett. **65**, 1820 (1990)

8.44   F. Gygi, M. Schluter: Phys. Rev. **B41**, 822 (1990)

8.45   R. Dittrich, C. Heiden: J. Vac. Sci. Technol. **A6**, 263 (1988)

8.46   R. Berthe, U. Hartmann, C. Heiden: preprint

8.47   M. Yoshimurka, K. Fujita, N. Ara, M. Kageshima, R. Shioda, A. Kawazu, H. Shigekawa, S. Hyodo: J. Vac. Sci. Technol. **A8**, 488 (1989)

8.48   C. Dai, C. Dai, C. Zhu, Z. Chen, G. Huang, X. Wu, D. Zhu, J.D. Baldeschwieler: J. Vac. Sci. Technol. **A8**, 484 (1989)

8.49   H. Bando, S. Kashiwaya, H. Tokumoto, H. Anzai, N. Kinoshita, K. Kajimura: J. Vac. Sci. Technol. **A8**, 479 (1990)

8.50   R. Fainchstein, J.C. Murphy: J. Vac. Sci. Technol., to be published

8.51   S. Vieira, M.A. Ramos, A. Buendia, A.M. Baro: Physica C **153-155**, 1004 (1988)

8.52   D. Anselmetti, H. Heinzelmann, R. Wiesendanger, H. Jenny, H.-J. Güntherodt, M. Düggelin, R. Guggenheim: Physica B – C **153-155**, 1000 (1988)

8.53   J.K. Grepstad, Ph. Niedermann, J.-M. Trescone, L. Antognazza, M.G. Karkut, O. Fisher: Physica C **153-155**, 1453 (1988)

8.54   A.M. Okoniewski, J.E. Klemberg-Sapieha, A. Yelon: Appl. Phys. Lett. **53**, 152 (1988)

8.55   M.D. Kirk, C.B. Eom, B. Oh, S.R. Spielman, M.R. Beasley, A. Kapitulnik, T.H. Geballe, C.F. Quate: Appl. Phys. Lett. **52**, 2072 (1988)

8.56   S.L. Pryadkin, V.S. Tsoi: Pis'ma Zh. Eksp. Teor. Fiz. **49**, 268 (1989) [JETP. Lett. **49**, 305 (1989)]

8.57   I.B. Al'tfeder, A.P. Volodin, I.N. Makarenko, S.M. Stivhoy, A.V. Shubnikov: Pis'ma Zh. Eksp. Theor. Fiz. **50**, 458 (1989) [JETP Lett. **50**, 490 (1989)]

8.58   X.L. Wu, Z. Zhang, Y.L. Wang, C.M. Lieber: Science **248**, 1211 (1990)

8.59   Z. Zhang, C.M. Lieber: to be published in J. Vac. Sci. Technol.

8.60   N.J. Zheng, U. Knipping, I.S.T. Tsong, W.T. Petuskey, J.C. Barry: J. Vac. Sci. Technol. **A6**, 457 (1988)

8.61   Ph. Niedermann, O. Fisher: J. Microsc. **152**, 93 (1988)

8.62   M.C. Gallagher, J.G. Adler: Physica C **162-164**, 1129 (1989)

8.63   M.D. Kirk, J. Nogami, A.A. Baski, D.B. Mitzi, A. Kapitulnik, T.H. Geballe, C.F. Quate: Science **242**, 1673 (1988)
8.64   L.E.C. van de Leemput, P.J.M. van Bentum, F.A.J.M. Driessen, J.W. Gerritsen, H. van Kempen, L.W.M. Schreurs, P. Bennema: J. Microsc. **152**, 103 (1988)
8.65   H. Heinzelmann, D. Anselmetti, R. Wiesendanger, H.-J. Güntherodt, E. Kaldis, A. Wisard: Appl. Phys. Lett. **53**, 2447 (1988)
8.66   T. Nakamura, T. Kimura, T. Kimura, M. Ihara, I. Umebu: J. Vac. Sci. Technol. **A8**, 472 (1990)
8.67   L.E.C. van de Leemput, P.J.M. van Bentum, F.A.J.M. Driessen, J.W. Gerritsen, H. van Kempen, L.W.M. Schreurs, P. Bennema: J. Crystal Growth, **98**, 551 (1989)
8.68   P. Hartman: in An introduction to crystal growth. P. Hartman, ed. (North Holland, Amsterdam 1973) p. 367
8.69   P. Niedermann, H.J. Scheel, W. Sadowski: J. Appl. Phys. **65**, 3274 (1989)
8.70   H.J. Scheel, P. Niedermann: J. Crystal Growth **94**, 281 (1989)
8.71   A.L. de Lozanne: Proc. Tsukuba Seminar on High T_c Superconductivity, Tsukuba, Japan 1989, p. 161
8.72   P. Muralt, D.W. Pohl: Appl. Phys. Lett. **48**, 514 (1986)
8.73   J.R. Kirtley, S. Washburn, M.J. Brady: Phys. Rev. Lett. **60**, 1546 (1988)
8.74   A.D. Kent, I. Maggio-Aprile, Ph. Niedermann, Ch. Renner, J.-M. Triscone, M.G. Karkut, L. Amtognazza, O. Fisher: Physica C **162–164**, 1035 (1989)
8.75   A.D. Kent, I. Maggio-Aprile, Ph. Niedermann, Ch. Renner, O. Fisher: J. Vac. Sci. Technol. A **8**, 459 (1990)
8.76   A.D. Kent, I. Maggio-Aprile, Ph. Niedermann, O. Fisher: Phys. Rev. B **39**, 12363 (1989)
8.77   X.L. Wu, C.M. Lieber, D.S. Ginley, R.J. Baughman: Appl. Phys. Lett. **55**, 2129 (1989)
8.78   D.G. Anselmetti: Thesis, University of Basel (1990)
8.79   L.L. Soethout: Thesis University of Nijmegen (1990)
8.80   I.B. Al'tfeder, A.P. Volodin, V.A. Grazhulis, A.M. Ionov, S.G. Karabashev: Pis'ma Zh. Eksp. Theor. Fiz. **50**, 182 (1989) [JETP Lett. **50**, 204 (1989)]
8.81   C.K. Shih, R.M. Feenstra, J.R. Kirtley, G.V. Chandrashekhar: Phys. Rev. B **40**, 2682 (1989)
8.82   see for example N. Nücker, H. Romberg, X.X. Xi, J. Fink, B. Gegenhei, Z.X. Zhao: Phys. Rev. B **39**, 6619 (1989)
8.83   see for example T. Takahashi, H. Katayama, Y. Okabe, S. Hosoya, K. Seki, H. Fujimoto, M. Sato, H. Inokuchi: Nature **334**, 691 (1988)
8.84   M. Tanaka, T. Takahashi, H. Katayama-Yoshida, S. Yamazaki, M. Fujinami, Y. Okabe, W. Mizutani, M. Ono, K. Kajimura: Nature **339**, 691 (1989)
8.85   M. Tanaka, S. Yamazaki, M. Fujinami, T. Takahashi, H. Katayama-Yoshida, W. Mizutani, K. Kajimura, M. Ono: J. Vac. Sci. Technol. **A8**, 475 (1990)
8.86   T. Hasegawa, M. Nantoh, H. Suzuki, N. Motohira, K. Kishio, K. Kitazawa: Physica **B165/166**, 1563 (1990)
8.87   Z. Zhang, Y.L. Wang, X.L. Wu, J.-L. Huang, C.M. Lieber: Phys. Rev. **B42**, 1082 (1990)
8.88   C.J. Chen, C.C. Tsuei: Solid State Commun. **71**, 33 (1989)
8.89   H.R. Flensberg, P. Hedegard, M. Brix: Phys. Rev. **B38**, 841 (1988)
8.90   P.W. Anderson, Z. Zou: Phys. Rev. Lett. **60**, 132 (1988)
8.91   see for example J.F. Zasadsinski, N. Tralshawala, D.G. Hinks, B. Dabrowski, A. Mitchell, D.R. Richards: Physica C**158**, 519 (1989)
8.92   J.R. Kirtley, C.C. Tsuie, S.I. Park, C.C. Chi, J. Rozen, M.W. Shafer: Phys. Rev. B**35**, 7216 (1987)
8.93   P.J.M. van Bentum, H.F.C. Hoevers, H. van Kempen, L.E.C. van de Leemput, M.J.M.F. de Nivelle, L.W.M. Schreurs, R.T.M. Smokers, P.A.A. Teunissen: Physica C **153-155**, 1718 (1988)
8.94   J.R. Kirtley, C.C. Tsuei, S.I. Park, C.C. Chi, J. Rozen, M.W. Schafer: Phys. Rev. **35**, 7216 (1987)
8.95   J.R. Kirtley, D.J. Scalapino: To be published
8.96   N. Bulut, D. Hone, D.J. Scalapino, N.E. Bickers: Phys. Rev. B**41**, 1797 (1990)
8.97   A. Millis, H. Monier, D. Pines: To be published

8.98   S.L. Cooper, M.V. Klein, B.G. Pazol, T.P. Rice, D.M. Ginzburg: Phys. Rev. **B37**, 5920 (1988)

8.99   P.J.M. van Bentum, G. Martinez, P. Lejay: Physica C **162-164**, 1237 (1989)

8.100   C.M. Varma, P.B. Littlewood, S. Schmitt-Rink, E. Abrahams, A.E. Ruckenstein: Phys. Rev. Lett. **63**, 1996 (1989)

8.101   P.J.M. van Bentum, L.E.C. van de Leemput, L.W.M. Schreurs, P.A.A. Teunissen, H. van Kempen: Phys. Rev. **B36**, 843 (1987)

8.102   J.-C. Wan, A.M. Goldman, J. Maps: Physica C **153-155**, 1377 (1988)

8.103   R. Wilkins, M. Amman, R.E. Soltis, E. Ben-Jacob, R.C. Jaklevic: Phys. Rev. **B41**, 8904 (1990)

8.104   M.A. Ramos, S. Vieira: Physica C **162-164**, 1045 (1989)

8.105   S. Vieira, M.A. Ramos, M. Vallet-Regi, J.M. Gonzalez-Calbet: Phys. Rev. **B38**, 9295 (1988)

8.106   J.R. Kirtley, R.M. Feenstra, A.P. Fein, S.I. Raider, W.J. Gallagher, R. Sandstrom, T. Dinger, M.W. Shafer, R. Koch, R. Laibowitz, B. Bumble: J. Vac. Sci. Technol. **A6**, 259 (1988)

8.107   J. Geerk, X.X. Xi, G. Linker: Z. Phys. B – Cond. Mat. **73**, 329 (1988)

8.108   J. Geerk, G. Linker, G. Meyer, Q. Li, R.-L. Wang, X.X. Xi: Physica C **162-164**, 837 (1989)

8.109   M. Gurvitch, J.J. Valles Jr., A.M. Cucolo, R.C. Dynes, J.G. Garno, L.F. Schneemeyer, J.V. Wasczak: Phys. Rev. Lett. **63**, 1008 (1989)

8.110   S. Vieira, J.G. Rodrigo, M.A. Ramos, N. Agrait, K.V. Rao, Y. Makino, J.L. Costa: J. Appl. Phys. **67**, 5026 (1990)

8.111   S. Vieira, J.G. Rodrigo, M.A. Ramos, K.V. Rao, Y. Makino: Phys. Rev. **B40**, 11403 (1989)

8.112   M.C. Gallagher, J.G. Adler: J. Microsc. **152**, 123 (1988)

8.113   M.C. Gallagher, J.G. Adler, J. Jung, J.P. Franck: Phys. Rev. **B37**, 7846 (1988)

8.114   M. Tachiki, S. Takahashi, F. Steglich, H. Adrian: Z. Phys. – Condensed Matter **80**, 161 (1990)

8.115   J.C. Phillips: Phys. Rev. **B41**, 8968 (1990)

8.116   R.C. Dynes, J.P. Garno, G.B. Hertel, T.P. Orlando: Phys. Rev. Lett. **53**, 2437 (1984)

8.117   A.P. Fein, J.R. Kirtley, M.W. Shafer: Phys. Rev. **B37**, 9738 (1988)

8.118   A.P. Volodin, M.S. Khaikin: Sov. Phys. JETP Lett. **46**, 508 (1989)

8.119   J.S. Tsai, I. Takeuchi, J. Fujitu, T. Yoshitake, S. Miura, S. Tanaka, T. Terashima, Y. Bando, K. Lijima, K. Yamamoto: Physica C **153-155**, 1385 (1988)

8.120   I. Igushi, H. Watanabe, Y. Kasai, T. Mochiku, A. Sugishita, E. Yamak: Physica **B148**, 322 (1987)

8.121   Mark Lee, D.B. Mitzi, A. Kapitulnik, M.R. Beasley: Phys. Rev. **B39**, 801 (1989)

8.122   A. Edgar, C.J. Adkins, S.J. Chandler: J. Phys. C**20**, L1009 (1987)

8.123   B.A. Aminov, L.I. Leoniuk, K. Sethupathi, M.V. Sudakova, Ya.G. Ponomarev, M.V. Pedyash, D.K. Petrov, H. Rahimov, L. Rosta: To be published

9. Recent Developments

As STM instrumentation has advanced, this novel powerful local probe technique has found even more widespread applications and much progress has been made, particularly in studies of adsorbates and molecular systems. In the following we shall review some of the most interesting recent developments in the field covered in this volume.

9.1 STM on Metal Surfaces[1]

In the last 2 to 3 years, studies of absorbate and molecular systems, as well as their dynamic properties have drawn more attention than have investigations of equilibrium structures of clean surfaces. Since adsorbates and molecules on metal surfaces will be covered in Sects.9.2 and 3, recent developments on clean metal surfaces will only be dealt with in this section.

One of the puzzling STM results on metal surfaces has been the large corrugation amplitudes with atomic resolution on close-packed surfaces, such as Au(111), Al(111) and Cu(111), as shown in Fig.3.11 [3.43-46]. Tunneling spectroscopies of I-V and I-s (s being the tunneling gap distance) were analyzed on Cu(111), Ag(111) surfaces when anomalous corrugations were observed [9.1]. As depicted in Fig.9.1, three different I-s curves were obtained with the same current and three different voltages. Three curves share the same slope (the same barrier heights, $d(\ln I)/ds$) with the ordinary tunneling gap, but they reveal three different slopes with small gaps. This result was explained by the presence of a foreign atom at the apex of the tunneling tip with an unusual electronic state. Abnormal I-s relations were reported on corrugated surfaces, such as Au(110)-(1×2) and Cu(110). It was suggested that the local barrier heights derived from $d(\ln I)/ds$ measurements may not truly represent those proposed by the theory for corrugated surfaces.

Surface melting behavior on Ga(001), (111) and (110) surfaces was observed [9.2]. The surfaces are stable up to the bulk melting temperature. At just below the melting temperature, the surfaces remain flat but micrometer-size hill-rock structure sets in. On the Ga(110), orientation-

---

[1] This section is based on notes provided by Y. Kuk.

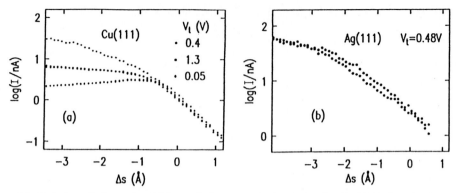

Fig.9.1. I-s relation when high resolution was obtained. (**a**) On Cu(111), I = 1 nA at three different biases. (2) On Ag(111), I = 2 nA was used. Figure from [9.1]

dependent disordering was observed, suggesting a specific softening mode as the surface melts. Further study of this kind may lead to an understanding of the melting mechanism. Surface diffusion and step-step interaction were investigated on Cu(11,1,1) and (810) surfaces [9.3]. The calculated surface diffusion coefficient is on the order of 10^{-16} cm^2/s. The step-step interaction mechanism and activation energy were estimated by the step movement.

Very recently, standing waves created by the scattering of surface-state electrons near step edges and foreign atoms were reported by two groups [9.4,5]. The surface states within the bulk band gaps on metallic surfaces [Au(111) and Cu(111)] may confine the electron scattering waves on the surfaces, resulting in quantum-mechanical interference. The interference between the incident and the reflected surface waves by steps and defects were observed in the images of $d(\ln I)/d(\ln V)$ generated from the position-sensitive I-V measurements. The period of the standing wave changes with the tunneling voltage due to the energy dispersion of the surface state (Fig.9.2). The scattering mechanism including the intensity and phase shift as a function of the surface-state energy was studied by this technique. Model quantum-mechanical boxes and wells in 2-D were demonstrated by such studies.

For metal alloys, it was shown that a distinction between different chemical species appears to be feasible under special tunneling conditions related with the microscopic structure of the probe tip [9.6]. For instance, an adsorbed atom at the apex of the tip may tend to form a chemical bond more likely with one element than with another, which can cause a difference in corrugation amplitudes in constant-current STM traces over the two different chemical species.

However, additional information about the tip structure is required in order to attribute the "higher" sites to a particular component of the metal alloy.

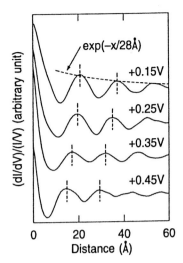

Fig.9.2. Distance dependence of d(lnI)/d(lnV) at four different bias voltages from a <112> step. Different dispersion and phase shifts result in the change of periods and positions. Figure from [9.4]

9.2 Adsorption on Metal Surfaces[2]

This section gives an overview of recent STM investigations on the adsorption of atoms and small molecules on well-defined, single-crystal metal surfaces - one of the major fields of surface science. As was the case at the time of the first edition of this quasi-monograph three years ago, most of the work has been done on structures of adsorbate layers; the investigations are listed in tabular form. In addition, some qualitatively new developments will be outlined: The dynamics of surface processes, the chemical reactions on surfaces that involve reconstructions, the phase transitions in the adsorbate layer, and the metal-on-metal epitaxy.

9.2.1 Dynamics of Surface Processes

Because STM is based on the mechanic movement of macroscopic parts (piezo ceramics, tip and holder), it is a relatively slow method, at least, in the constant-current mode. Nevertheless it was successfully applied to the study of dynamic processes on surfaces. One example is surface diffusion. Diffusion constants span an extremely wide range of orders of magnitude, so that in some favorable situations the adsorbates move at a speed that is detectable by STM (most experimental set-ups are still restricted to room-temperature measurements). This was achieved for S atoms adsorbed on a Re(0001) surface [9.7,8]. The S atoms form a lattice gas between ordered (2×2) islands; from the spatial correlation function of the disordered area,

[2] This section is based on notes provided by J. Wintterlin, except for Sect.9.2.5.

the diffusion energy could be extracted. Another example is the finding of "hot" atoms that occur during the dissociation of oxygen molecules on an Al(111) surface [9.9]. The existence of such non-thermal processes during adsorption had been discussed before. While the "hot" particles themselves were not detected (a lifetime of 1 ps has been estimated), an analysis of the distribution of oxygen atoms after they became equilibrated provided clear evidence for such a mechanism. Figure 9.3a displays an STM topograph of the Al(111) surface recorded after adsorption of a very small amount of oxygen (Θ_{ox} = 0.0014). The small dark spots represent adsorbed oxygen atoms (oxygen is imaged as a depression at the tunnel current chosen in the experiment), the two horizontal lines are atomic steps of the aluminum. It was found that each of the dark spots, in fact, represents a single oxygen atom. First pairs of atoms, on neighboring (1×1) sites, such as the one visible in the inset, were found only at higher exposures. (The image shown in the inset was recorded with a higher tunnel current by which a central bright spot becomes visible in each of the oxygen features). The finding of individual atoms is surprising, since it means that the distribution of oxygen has no "memory" of the dissociation from the oxygen molecules. One would expect to see pairs of atoms on closely neighboring sites. Figure 9.3b quantifies this impression that pairs are absent: it shows the density of oxygen atoms in Fig.9.3a around a selected atom as a function of the distance from this center atom (averaged over all atoms in Fig.9.3a as central atoms and normalized to the mean density). Obviously within the scatter of the data for small separations (the number of data points increases with distance) the distribution is random, i.e., there is no correlation between the positions of the atoms. This cannot be explained by thermal motion of the adsorbed atoms as these atoms showed no displacements over time intervals of up to 1 h. It was therefore concluded that the (atomic) adsorption energy connected with the dissociation of the O_2 molecules is not directly released to the heat bath of the solid, but creates two "hot" oxygen atoms which fly apart by at least 80 Å before they get equilibrated. Such a mechanism can also explain the unusual growth behavior of the islands observed at somewhat higher exposures. (Chemisorbed oxygen forms (1×1) islands on Al(111) [9.10]). Figure 9.3c shows that between 3 and 13 Langmuir (L) of oxygen the size distribution changed only very little and most of the islands still consisted of merely one atom, although the coverage increased by a factor of four; for 20 L one could first observe a qualitative change. This is fully consistent with the "hot" atom picture: islands are formed when a "hot" atoms hits another one that is already equilibrated to which it can release its energy; because the masses are identical this should be a very efficient process. Clearly the probability for such an event is very small at the beginning when the coverage is small, and most atoms equilibrate on the bare surface. When the number of individual atoms increases, islands start to form. In this picture the islands grow by a non-thermal process, and it was suggested that it also affects the formation of oxide nuclei, which were observed at steps and (1×1) island edges in a

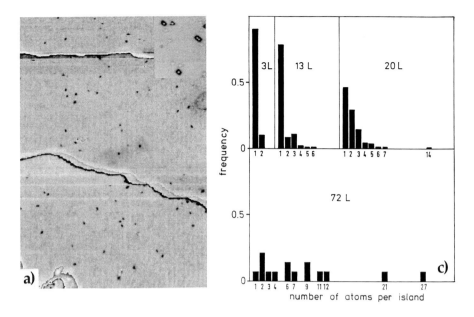

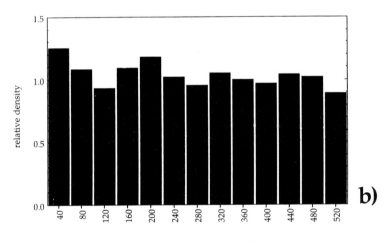

distance between adatoms [Å]

Fig.9.3. (a) STM topograph of an Al(111) surface after adsorption of a very small amount of oxygen. Dark spots represent oxygen atoms horizontal stripes, atomic steps. (I= 1nA, V = 0.2V, $474 \times 703 \text{Å}^2$). Inset: three single oxygen atoms and one pair, seen after a larger oxygen dose. (I = 8nA, V = 0.5V, $54 \times 63 \text{Å}^2$). (b) Normalized density of oxygen atoms as a function of separation between atoms. (c) Size distribution of oxygen (1×1) islands as a function of exposure. From [9.9]

later stage of the adsorption [9.10]. A similar explanation has been given for the observation that after Xe adsorption on a 4 K surface nearly all of the Xe atoms were found at the steps, and for higher coverages in small clusters, although the thermal mobility is negligible at this temperature [9.11]. It was concluded that the Xe atoms must scatter across the surface by hundreds of Ångstroms before becoming accommodated.

9.2.2 Adsorbate Structures

Table 9.1 lists adsorbate structures that have been studied since 1991. Only those structures that were prepared to adsorption from the gas phase were included, not those from solution. A considerable amount of work has been done to understand the imaging of adsorbed sulfur layers by modeling STM contours [9.12-14]. Calculations were also performed for adsorbed Xe atoms which showed convincing agreement with experimental images [9.15].

9.2.3 Adsorbate-Induced Reconstructions

The majority of papers which have been published since 1991 and treated adsorption processes on metal surfaces, dealt with adsorbate-induced reconstructions (Table 9.2). One of the reasons for this interest is that reconstructions are phenomena that occur quite frequently during adsorption and have occupied surface science from the very early days. They reflect the inherent geometric instability of clean metal surfaces which, apart from vertical relaxations, represent "truncated bulk" surfaces in their majority. This geometry is frequently lost when a system can gain an improved bonding to an adsorbate particle, e.g., by providing a higher coordination number. Because substrate atoms are involved in the adlayer, the structures are mostly more complex than those of simple adsorbate overlayers, and models were often disputed for years. STM could resolve some of these structure problems mostly, however, not directly from the topographs. Because the local density of states of these systems is naturally more complex than that of simple adlayers, a direct link of topographic features with atomic positions is often doubtful. Tip effects are, similar to the non-reconstructed adsorbate layers, a further complication, as was demonstrated for the $(2 \times 1)O$ added row reconstructions on Cu and Ni(110) [9.64]. With a clean W tip the oxygen atoms are imaged as protrusions, when an oxygen is adsorbed on the tip (by dosing or by transfer from the sample) the metal atoms are seen as protrusions. STM studies on reconstructions therefore used more indirect arguments about substrate mass transfer during the reconstruction (as shown in Chap.4 this can reveal the number of substrate atoms in the reconstructed unit cell), about the spatial relation of particular features in the STM contours with respect to the substrate atoms or to

Table 9.1. Adsorbate overlayers

Adsorbates/Surfaces	Structures	References
ethylidyne/Pt(111)	(2×2)	9.16,17
graphite/Pt(111)	"incommensurate"	9.18
graphite/Ni(111)	-	9.19
CO/Pt(111)	single molecules and clusters	9.20
O/Al(111)	(1×1)	9.9,10
O/Co(10$\bar{1}$0)	c(2×4), (2×1)$_{\theta=1}$	9.21
S/Pd(111)	($\sqrt{3}\times\sqrt{3}$)R30°	9.22
S/Re(0001)	(2×2), c($\sqrt{3}\times5$)rect, $(3\sqrt{3}\times3\sqrt{3})$R30°, $\begin{vmatrix} 3 & 1 \\ 1 & 3 \end{vmatrix}$	9.7,8,12-14, 23,24
S/Ru(0001)	(2×2), ($\sqrt{3}\times\sqrt{3}$)R30°	9.25
S/Ni(100)	(2×2), c(2×2)	9.26
S/Pd(100)	c(2×2)	9.27
S/Pt(100)	c(2×2)	9.28
S/Mo(100)	c(2×2), $\begin{vmatrix} 2 & \bar{1} \\ 1 & 1 \end{vmatrix}$, c(4×2), (2×1)	9.29
Cl/Cu(111)	($\sqrt{3}\times\sqrt{3}$)R30°, $(6\sqrt{3}\times6\sqrt{3})$R30°, "(4$\sqrt{7}$×4$\sqrt{7}$)R19°"	9.30
I/Pt(111)	(3×3)	9.31
I/Au(111)	(5×$\sqrt{3}$), (7×7)R22°	9.32
I/Pt	($\sqrt{2}\times5\sqrt{2}$)R45°	9.31,33,34
Xe/Pt(111)	single atoms and clusters	9.11
Xe/Ni(110)	single atoms	9.20,35
FeO/Pt(111)	"incommensurate"	9.36
(Cs+O/Ru(0001)	($\sqrt{7}\times\sqrt{7}$)R19°	9.37

another known structure [9.55], or about particle mobility, which is different from that expected for the unreconstructed case [9.39]. Not included in Table 9.2 are adsorbate-induced facetting which can be regarded as an extreme type of reconstruction. They were observed during the interaction of oxygen with a stepped Ag(110) surface [9.77] and during the oxidation of Ni(100) [9.78].

A recent development of interest are studies about surface-chemical reactions where one or several of the reacting particles induce a reconstruction. Systems studied were the reaction of H_2S with O/Ni(110)-(2×1) [9.79], the decomposition of NO on Rh(110) [9.80], the methanol

Table 9.2. Adsorbate-Induced Reconstructions

Adsorbates/Surfaces	Structures	References
H/Ni(110)	streaky (1×2)	9.38
Na/Al(111)	$(\sqrt{3}\times\sqrt{3})R30°$, (2×2)	9.39
K/Cu(110)	(1×3), (1×2)	9.40,41
Cs/Cu(110)	(1×4), (1×3), (1×2)	9.41.42
Na and K/Au(111)	substrate contraction	9.43
Na/Au(111)	$c(4\times2)$, "(13×13)"	9.44
K/Au(100)	(1×2)	9.45
K/Au(110)	(1×5), (1×3), (1×2), $c(2\times2)$	9.41,46
C/Ni(100)	$(2\times2)p4g$	9.47,48
C/Ni(110)$_{stepped}$	(4×1)	9.49
N/Ni(100)	$(2\times2)p4g$	9.50
N/Cu(100)	$c(2\times2)$	9.51
N/Ni(110)	(2×3)	9.52
N/Cu(110)	(2×3)	9.53-55
O/Cu(111)	$\begin{vmatrix} 4 & 3 \\ -3 & 5 \end{vmatrix}, \begin{vmatrix} 9 & 1 \\ 1 & 5 \end{vmatrix}$	9.56,57
O/Cu(100)	$(2\sqrt{2}\times\sqrt{2})R45°$	9.58,59
O/Ni(110)	(3×1), (2×1), (3×1), (9×5)	9.60
O/Ni(110)$_{stepped}$	(3×1), (2×1)	9.61-63
O/Cu(110)	(2×1)	9.64,65
O/Ag(110)	$(n\times1)$ $\{n=7,6,...,2\}$	9.66-69
O/Rh(110)	$(2\times2)p2mg$, $c(2\times6)$, $c(2\times8)$, $c(2\times10)$	9.70,71
O/Co(10$\bar{1}$0)	$(2\times1)_{\theta=0.5}$	9.21
O/W(100)	(2×1)	9.72
S/Al(111)	"incommensurate"	9.73
S/Ni	$(5\sqrt{3}\times2)$	9.74
S/Cu(111)	$\begin{vmatrix} 4 & 1 \\ -1 & 4 \end{vmatrix}$, $(\sqrt{7}\times\sqrt{7})R19°$	9.30,75
S/Pd(111)	$(\sqrt{7}\times\sqrt{7})R19°$	9.22
S/Ni(110)	(4×1)	9.76

oxidation on O/Cu(110)-(2×1) [9.81,82], the CO oxidation on O/Rh (110)-$c(2\times2)$ [9.83], and the oxygen adsorption on alkali/Cu(110)-$(1\times n)$ [9.84]. An example that shall be discussed in more detail is the decomposition of ammonia on the Ni(110) surface (Fig.9.4) [9.85]. It had earlier been observed that small amounts of oxygen enhance the dissociation of ammonia, but it was difficult to understand why the reactivity showed a maximum at Θ_{ox} ~ 0.16 and dropped with higher oxygen coverages. In the

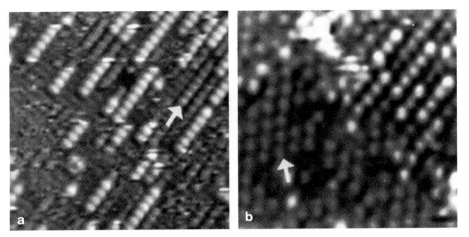

Fig.9.4. (a) STM image of Ni(110), precovered with $\Theta = 0.17$ of oxygen, after subsequent exposure to 0.2 L NH$_3$ ($90 \times 90 \text{Å}^2$). Bright chains are -Ni-O- added rows,
less bright features (arrow) rows of OH molecules. (b) Image after exposure of 5 L of
NH$_3$ ($55 \times 55 \text{Å}^2$). The arrow points to a c(2×2) structure formed by NH$_2$ molecules.
From [9.85]

presence of adsorbed oxygen the dissociation proceeds via: NH$_3$ + O$_{ad}$ →
NH$_{2,ad}$ + OH$_{ad}$; OH$_{ad}$ reacts further to H$_2$O which finally desorbs. It was
demonstrated by STM that oxygen on Ni($\bar{1}$10) induces a similar added-row
reconstruction as on Cu(110) (Chap.4) consisting of adsorbed chains of
nickel and oxygen atoms [9.60]. In Fig.9.4a these are visible as bright rows
of atoms, the maxima represent the Ni atoms, the minima the locations of
the O atoms. Figure 9.4a illustrates the situation after predosing with 0.17
monolayers of oxygen and adsorption of a small amount of ammonia. This
led to the appearance of new features, the less bright rows (see arrow)
which run along the same direction as the added rows. *Eierdal* et al. [9.60]
interpreted these as rows of hydroxyl from the reaction of oxygen with
hydrogen (no reconstruction). The NH$_{2,ad}$ fragments are not visible here
which, as it was suggested, was caused by their high mobility at low coverage. Only after higher ammonia doses did they become frozen in an ordered
structure and detected as a c(2×2) structure (see the arrow in Fig.9.4b).
Since the added row reconstruction is partially lifted by the reaction
because of the removal of oxygen, Ni atoms are also released and are seen
as new terraces on larger-scale images. The decisive observation which led
to an understanding of the reactivity maximum for a certain oxygen coverage, was the distribution of the added rows: as visible in Fig.9.4a at low
coverages they form more-or-less isolated, relatively short chains. When
only the end-standing oxygen atoms in the -Ni-O- chains can react, it is
clear that the reactivity will go down once the rows start to grow together
at higher coverages. In addition, at low coverages, the chains are able to
spontaneously fragment, by which new terminating O atoms are created.
These fragmentations are also reduced at higher coverages because the

neighboring sites necessary for this process are diminished. The example demonstrates how a surface-chemical reaction can be affected by the microscopic topography of the surface which is a direct consequence of a particualr reconstruction mechanism.

9.2.4 Phase Transitions

Another, more recent, application of STM has been the investigation of phase transitions which occur in some adsorbate systems. Phase equilibria between ordered islands and disordered phases, which mark first-order transitions, have been observed for Cs/oxygen compounds on a Ru(0001) surface [9.37] and for the above mentioned example of S on Re(0001) [9.7, 8]. Continuous, two-dimensional phase transitions have theoretically been studied extensively in the past; experiments were largely performed by diffraction techniques, which provide macroscopic parameters such as critical exponents. STM experiments are therefore of great interest in order to check predictions by theory on a microscopic basis, too. However, only a very few systems have been studied so far. An example for an adsorbate-on-metal system is the Cu(110) surface covered with alkali metals [9.40, 41]. As shown in the side view (Fig. 9.5c), the alkali atoms (large gray circles) cause $(1 \times n)$ missing row-type reconstructions of the Cu surface where close-packed rows of Cu atoms (along the $[1\bar{1}0]$ direction) are removed and the alkali metal atoms occupy sites in these troughs. The separations between the rows are determined by repulsive dipole-dipole interactions between the alkali atoms and, hence, by the alkali coverage, whereas the rows themselves are kept together by net attractive interactions caused by the reconstruction, giving rise to $(1 \times n)$ structures. Phase transitions of such uniaxial $(1 \times n)$ phases have been treated theoretically. It was predicted that a uniaxial (1×3) phase disorders by a two-step mechanism: in a first step by a transition from a commensurate solid into an incommensurate floating solid, which is characterized by a regular distribution of domain walls (algebraically decreasing density-density correlation) in a second step by transition into an incommensurate fluid which is characterized by an exponentially decreasing density-density correlation. For the alkali/Cu system the disordering was realized by decreasing the coverage of the alkali metals (Cs in the present example) from that of a perfect (1×3) structure ($\Theta_{Cs} = 0.13$) [9.42]. Figure 9.5a exhibits an STM image of the Cu(110) surface with $\Theta_{Cs} = 0.12$. The bright lines are the Cs-filled missing rows which form (1×3) domains, the darker streaks are domain walls which represent local (1×4) elements (see the model in Fig. 9.5c). Sudden horizontal cutoffs of the walls are due to thermal motion of the reconstruction pattern, which signals that the system is actually in thermodynamic equilibrium. From the regular sequence of domain walls it was concluded that this phase is in agreement with an incommensurate floating solid, as expected from theory. However, the finite width of the terraces prevented

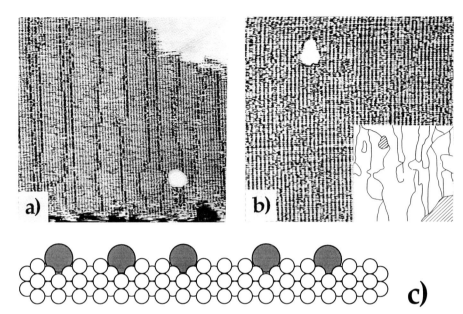

Fig.9.5. (a) STM image of Cu(110) with $\Theta = 0.12$ of Cs; (1×3) domains separated by (1×4) walls $(795 \times 800 \text{Å}^2)$. (b) $\Theta_{Cs} = 0.12$; (1×4) domains separated by disordered areas. The inset dislays the domain structure; defects and steps are hatched. The bright streaks are Cs-filled missing row troughs. (c) Model (side view) of the alkali metal induced (1×3) reconstruction of Cu(110) with one (1×4) domain boundary. Large gray circles are alkali atoms, small white circles the Cu atoms. From [9.42]

an evaluation of the decay of the correlation function from the STM images, which would be required for a definite conclusion about the phase. Figure 9.5b displays the surface after the Cs coverage was further reduced to 0.09. It shows (1×4) domains (indicated in the inset) which are separated by small disordered areas exhibiting (1×3) and (1×5) spacings. From the absence of a regular sequence of (1×4) domains *Schuster* et al. [9.42] concluded that there is no long-range order, indicating that this image belongs to the incommensurate fluid phase, also in agreement with predictions. It is clear that progress in this kind of studies will be greatly enhanced by the availability of variable-temperature STMs.

9.2.5 Metal-on-Metal Epitaxy[3]

STM studies of growth processes, in this case metal-on-metal epitaxy and related aspects, have thrived particularly over the last years. It was mentioned already in Sect.4.4 that this area has been studied intensively over the last decades, both experimentally and theoretically [9.86]. Various different

---

[3] This section is based on notes provided by R.J. Behm.

surface-sensitive methods, in particular diffraction techniques, were employed and a large body of experimental data was accumulated; on the other hand, well-funded theoretical concepts describing the thermodynamics and kinetics of epitaxial growth were developed (Sect.4.4) and [10.86]. A detailed understanding, however, suffered from the fact that film morphology and growth process were experimentally accessible only by indirect means. This is different for STM studies which give a direct, real-space image of the (local) film morphology and, hence, for the first time allow direct conclusions on the growth processes to be drawn. This new situation and the ongoing interest in metal thin-film growth, both from technical aspects as well as from basic scientific interest, have stimulated a large number of STM studies in this area.

The evolution of film morphology with increasing film thickness, under given growth conditions, is the primary interest in such studies. This has so far been investigated for a large number of metal epitaxial systems, ranging from simple growth systems, where the substrate can be considered as an inert base, to more complex systems where this is no longer the case, and substrate and deposited atoms can intermix in the interface region. In the latter case deposited atoms can exchange with substrate surface atoms, or even diffuse into the bulk (or vice versa), i.e., growth processes compete with exchange and interdiffusion processes. In addition, the often drastic modifications of the growth process caused by the presence of pre- or co-adsorbed species, well known as so-called *surfactant effects* [9.87], were investigated in several case studies. A (presumably incomplete) account of the different metal-on-metal epitaxial systems studied so far by STM is given in Table 9.3.

A general result of these studies is that upon deposition under "normal" conditions (300K sample temperature and ~1 monolayer per minute deposition rate) film growth does not proceed in the way predicted by the growth modes, which had been derived on the basis of thermodynamic considerations [9.88], but is strongly affected, or even dominated, by kinetic effects. It follows a scheme where initially small clusters of a few deposited adatoms are nucleated, they then grow by incorporation of migrating adatoms into two-dimensional islands of mostly monolayer thickness. These islands continue their lateral growth until they coalesce. Vertical growth sets in and proceeds by nucleation of second-layer islands on top of first-layer islands, and so forth. The overall morphology of the resulting film is then determined by the partial coverage of the nth layer required to allow nucleation of (n+1) layer islands, i.e., layer-by-layer-like growth results when next-layer islands are formed only after coalescence of the lower-layer islands, while in other cases more-or-less rough films result, with several layers exposed simultaneously. In most cases the thermodynamically expected film morphologies only develop upon deposition at or annealing to elevated temperatures. The details of the growth process, such as island densities and shapes, onset of higher-layer formation, etc. however, were found to vary considerably [9.89-73]. Likewise, the impact of preadsorbed

Table 9.3. Metal-on-metal epitaxial systems

Homoepitaxy

Ag(111) [9.89-91]	Au(111) [9.96,97]	Ni(100) [9.92,94,101]
Au(100) [9.92-94]	Fe(100) [9.98-100]	Pt(111) [9.102-104]
Au(110) [9.95]		

Heteroepitaxy

Rh/Ag(100) [9.105]	Co/Cu(100) [9.122,123]	Cu/Pd(110) [9.142,143]
Au/Ag(110) [9.106,107]	Fe/Cu(100) [9.121,124-131]	Ag/Pt(111) [9.142,144-146]
Cu/Ag(111) [9.108]	Mn/Cu(100) [9.132]	Co/Pt(111) [9.147,63]
Ni/Ag(111) [9.108]	Ag/Cu(111) [9.133]	Ag/Ru(0001) [9.149]
NiAu(100) [9.109]	Co/Cu(111) [9.134,135]	Au/Ru(0001) [9.92,150-153]
Ni/Au(110) [9.110]	Fe/Cu(111) [9.136]	Co/Ru(0001) [9.92,94,154]
Ag/Au(111) [9.111,112]	Cr/Fe(001) [9.137.138]	Cu/Ru(0001) [9.151,152, 154-156]
Co/Au(111) [9.113,114]	Ag/Ni(100) [9.139]	
Fe/Au(111) [9.115,116]	Au/Ni(100) [9.109]	Ni/Ru(0001) [9.108]
Ni/Au(111) [9.97,117,118]	Au/Ni(110) [9.140]	Pt/W(110) [9.157]
Rh/Au(111) [9.119]	Au/Ni(111) [9.141]	Pt/W(111) [9.157,158]
Au/Cu(100) [9.120,121]		

Surfactant/Adsorbate Effects

Ag/Ag(111)-Sb [9.160]	Pt/Pt(110)-O [9.159]	Cu/Ru(0001)-O [9.160]
Ni/Ni(100)-O [9.93,101]	Au/Ru(0001)-O [9.94,153]	

Structure, Strain Relaxation

Pt/Pt(111) [9.103]	Ag/Pt(111) [9.146]	Cu/Ru(0001) [9.155,156]
Co/Ru(0001) [9.154]	Ni/Ru(0001) [9.106]	

surface species on the growth characteristics differs largely between different systems [9.89,91,93,94,101,153,159,160]. Finally, in a large number of the systems listed in Table 9.3, exchange and even interdiffusion processes were observed. In one case, for Au/Ni(110), exchange could even be rationalized on the basis of effective-medium calculations [9.140]. Another phenomenon, surface etching during deposition, was explained to result from exchange processes [9.135].

A more detailed understanding of epitaxial growth was gained from studies focussing on the characterization and qualitative evaluation of the different processes contributing to the overall growth process. This includes nucleation (*critical cluster size*) and growth of two-dimensional islands, adatom migration on the flat terraces and along terrace and island edges (*intralayer transport*) as well as material transport between layers (*interlayer transport*), by adatoms crossing over descending steps (*island and terrace edges*). Intralayer transport determines the shape and density of adatom islands, while interlayer transport is responsible for the smoothness/roughness of the film. The relative rates of the respective processes, and hence their activation barriers, are decisive for kinetically controlled growth.

Because of the simpler energetics, these studies were predominantly performed for homoepitaxial systems, e.g., for Pt(111), Fe(100), Au(100) and Ag(111) homoepitaxy. For instance, for Pt(111) homoepitaxy it could be demonstrated how the island shape varies with deposition temperature, due to the varying contribution of the different edge mobilities [9.102]. Critical cluster sizes in Fe(100) and Au(100) homoepitaxy where determined from the flux and temperature dependence of the island densities [9.93, 100]. In the latter case it could even be demonstrated that self-diffusion on the "hex" reconstructed Au(100) surface is strongly anisotropic [9.93], namely much faster along the reconstruction rows than perpendicular to them, and energy barriers for adatom migration on the terrace could be derived. The effect of flux and deposition rate on island density and island shape in this system is evident from the two STM images in Fig. 9.6a,b, which show a Au(100) surface after deposition of 0.2 MonoLayers (ML) Au at different deposition conditions (Fig.9.6a: 0.5ML/min, Fig. 9.6b: 0.005ML/min; T = 315K). Not only is the island density lower by a factor of 5 in (Fig.9.6b), but also the aspect ratio of the anisotropic islands has grown from 3 at the higher flux to 8 at the lower flux, for more details see [9.93]. Due to adatom trapping at ascending steps there is a depletion area with no adatom islands along the steps on the lower terrace. The width of this area is determined by the mobility of the adatoms. The images in Fig.9.6c,d illustrate clearly that these areas have a different width for steps parallel to the reconstruction rows than for steps perpendicular to the rows, which is direct evidence for different adatom mobilities in the two directions. The additional depletion zone on the upper terrace side signals that the barrier for adatoms to pass the descending step is small or not existent in this case, which is the underlying reason for the observation of an almost perfect layer growth in this system. For Ag(111) homoepitaxy, where growth follows a multilayer growth behavior with many layers exposed from rather low coverages on [9.89-91], this barrier could be estimated from the number of layers exposed for different coverages [9.91]. It should be noted that it is not the absolute barrier at the step, but the ratio of this barrier to the barrier for adatom migration, which is decisive for the intralayer transport and therefore for the roughness of the resulting film.

The success of these quantitative studies depends highly, of course, on the simultaneous rapid development of theoretical concepts. In particular, the comparison with results from mean-field rate-equation theory [9.161] and kinetic Monte-Carlo studies proved to be invaluable for the understanding of the experimental data [9.162].

Finally, we want to mention recent work in an area which is closely related to metal-on-metal epitaxy, namely epitaxial growth of ultra-thin oxide layers on metal supports. This is performed either by metal deposition in an O_2 atmosphere, or by subsequent oxidation of a thin deposited metal

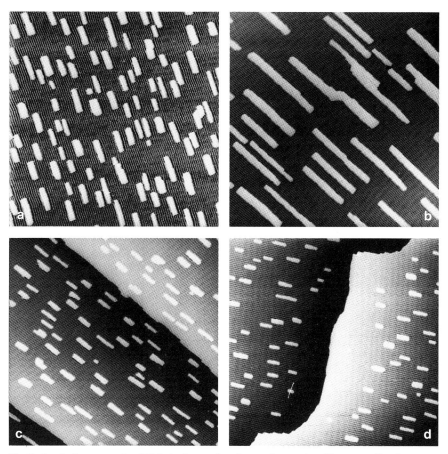

Fig.9.6a-d. Large-scale STM images of submonolayer Au films on "hex" reconstructed Au(100) ($150 \times 150 nm^2$). (**a**) and (**b**) present the variation in island density and island shape at different deposition fluxes but constant temperature and coverage, obtained with a flux of 0.5 Ml/min (**a**) and 0.0005 ML/min (**b**), respectively (T = 315 K, Θ = 0.2ML). (**c**) and (**d**) exhibit the different widths of the denuded zones along step edges parallel (**c**) and perpendicular (**d**) to the direction of the reconstruction rows (Θ = 0.5) [9.93]

film. In both cases the substrate has to be sufficiently inert against bulk oxidation to achieve and maintain a well-defined interface. First studies in this direction were therefore performed either on noble metals - Ni(100)/ Au(111) [9.163, 164]; Ni(100) and NiO(111) on Ag(111) [9.165] or on Pt [FeO/Pt(111) [9.166] but also on Rh(TiO$_x$/Rh(111) [9.167]. Although structure and growth of these films is rather complex, these studies gave a new insight in particular into the growth mechanism. For instance, NiO grows into well-defined two-dimensional islands on a Ag(111) substrate at temperatures around 500 K. It depends on the deposition rate and the O$_2$

partial pressure, which of the two orientations, NiO(100) or NiO(111), is formed. Over a range of flux/pressure conditions, islands of the two phases coexist on the surface [9.165].

Work in this area is still sparse at the moment, but is expected to expand rapidly, stimulated by the interest in thin-oxide layers, e.g., as model systems for catalysis studies.

9.3 Molecular Imaging by STM[4]

The Scanning Tunneling Microscope (STM) has recently been applied to many molecular systems in ultrahigh vacuum. Recent results on naphthalene, azulene, methylazulenes, benzene and ethylene on Pt(111), as well as C_{60} on several substrates, have been discussed. Progress has been made on distinguishing isomers on a surface, identifying molecular adsorption sites from STM images, calculating expected STM images for molecular systems, observing a surface chemical reaction, and observing molecular organization and the internal structure of C_{60} fullerenes.

9.3.1 STM of Molecules on Metals in Ultrahigh Vacuum

The STM imaging of small molecules adsorbed onto metal surfaces in ultrahigh vacuum has recently been performed on several molecule-substrate systems. Many studies have used Pt(111) as a substrate because of its chemical reactivity and applications to catalysis. STM images have now been used to distinguish two sets of isomers (naphthalene and azulene; and monomethylazulenes) on Pt(111) [9.168-171]. Benzene on Pt(111) was observed to have three distinct types of images for differing adsorption sites [9.172]. Remarkable progress has been made in calculating expected STM images for molecular systems, both by extended Hückel molecular-orbital calculations for the molecule-substrate system [9.169] and by an electron-scattering quantum-chemistry technique [9.173-175]. The dehydrogenation of ethylene to ethylidyne and then to graphite on Pt(111) has been observed directly in STM images [9.176-178]. Finally, many studies have been performed on the carbon fullerene C_{60}, on both metal [9.179-185] and semiconductor [9.186-189] substrates, to examine the molecular organization on the surface and the variation in STM images of the internal structure of the molecule, depending on adsorption site, molecular orientation, and tunneling bias voltage.

---

[4] This section is based on notes provided by S. Chiang.

9.3.2 Naphthalene, Azulene and Methylazulenes on Pt(111)

The molecular orientation and organization of naphthalene on Pt(111) has previously been investigated [9.190]. Recently, STM images of a series of seven related molecules have been measured on Pt(111) [9.168-171]. Theoretical calculations of the expected images showed good agreement with the experimental results [9.169]. The list of molecules analysed comprises naphthalene, its isomer azulene, 1-methylazulene (1-MA), 2-methylazulene (2-MA), 6-methylazulene (6-MA), 4,8-dimethylazulene (DMA), and 4,6,8-trimethylazulene (TMA) (Fig.9.7).

While naphthalene adsorbed on Pt(111) is typically observed as a bi-lobed structure with three distinct azimuthal orientations in STM images, occasional high-resolution images with particular tunneling tips are able to resolve a double-ring structure [9.168]. In contrast, its isomer azulene has a sticking coefficient on Pt(111) approximately one quarter that of naphthalene, as determined from observed coverage in STM images. Low-coverage azulene [≈ 0.2 MonoLayer (ML)] images show very few distinct features, as the molecular motion appears to be comparable to the line scan speed of 2.5 s. Increasing coverage to 1 ML hinders the molecular motion and permits resolution of individual molecules in close-packed domains with

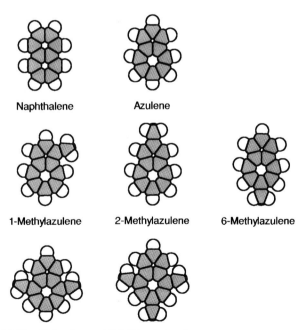

Naphthalene   Azulene

1-Methylazulene   2-Methylazulene   6-Methylazulene

4,8-Dimethylazulene   4,6,8-Trimethylazulene

Fig.9.7. Space-filling models showing structure of related molecules measured on Pt(111) in [9.168-171]. The molecules are all shown to scale, and the Van der Waals length of the naphthalene molecule is 8.1 Å. From [9.171]

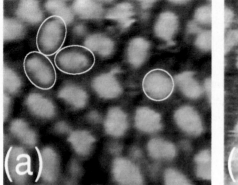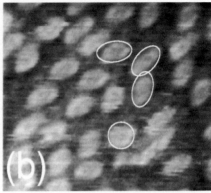

Fig. 9.8. Coadsorbed naphthalene and azulene with all three orientations of the elongated naphthalene molecules marked with ovals and symmetric azulene molecules marked by circles ($\approx 90 \times 90\,\text{Å}^2$; $-0.05\,\text{V}$; $1\,\text{nA}$). (**a**) Low resolution and (**b**) higher resolution with double rings apparent for naphthalene and singles rings observed for azulene. From [9.168]

(3×3) packing. Azulene molecules typically appear round in these room temperature images, with no appreciable asymmetry, perhaps as a result of rapid rotation of azulene molecules within individual adsorption sites [9.168].

Imaging coadsorbed molecules in mixed monolayers eliminates effects of the tunneling tip on the observed resolution of the molecules. The ability to recognize different molecules in the same image is also important for employing the STM to monitor surface reactions. Figure 9.8 displays two images of coadsorbed naphthalene and azulene on Pt(111) [9.168]. Imaging of low coverage of azulene is possible in this case because mobility of the azulene molecules on the surface is limited by the coadsorbed naphthalene. In Fig.9.8a the naphthalene molecules are noticeably elongated, while the azulenes appear circular and symmetric. The higher-resolution image in Fig.9.8b shows the double-ring structure of naphthalene with all three azimuthal orientations, indicating that the molecular asymmetry is not an artifact relating to the tunneling tip, while the azulene molecules appear to have only a single ring.

Images of coadsorbed layers of 1-MA and 2-MA indicate that these molecules can also be distinguished on the surface by their shapes, with 1-MA having a kidney-bean shape and 2-MA having a pear shape (Fig. 9.9d) [9.168, 169]. Images of coadsorbed naphthalene and 2-MA reveal that the long axis of 2-MA is rotated 30° with respect to that of naphthalene, and therefore also to the close-packed direction of the underlying Pt lattice. 6-MA also has a distinctive shape, having a diamond shape in low-resolution images, while the hole in the seven-carbon ring is resolved in high-resolution images (Fig.9.9e). DMA and TMA both appear as fourfold clover-leaf structures in low-resolution images, with a single bright spot

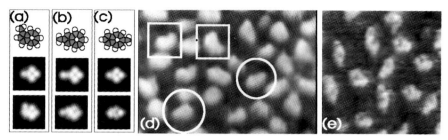

Fig.9.9a-e. Monomethylazulenes on Pt(111). Schematic, LUMO for isolated molecules (*upper plot*) and ρH (*lower plot*) for (**a**) 1-methylazulene (1-MA), (**b**) 2-MA, and (**c**) 6-MA. Plots of 1-MA and 2-MA are at 2 Å height, while that for 6-MA is at 0.5 Å height. (**d**) Low-resolution STM image of mixed 1-MA, ≈ 20% coverage (in squares, e.g.) and 2-MA (in circles, e.g.) near saturation coverage. Some molecules appear to be impurity naphthalene or azulene in motion. (**e**) High-resolution STM image of 6-MA. From [9.171]

appearing on one lobe of the molecule in many images. By comparison with the orientation of naphthalene molecules in mixed overlayers, the long axis of TMA is also aligned with [1$\bar{1}$0]. The relative sticking coefficients and translational diffusion rates for the series of molecules have also been inferred by comparing the observed molecular coverage in the images with the molecular exposure [9.169, 171].

A calculational technique based on extended Hückel theory for the molecule-substrate system has been developed for simulating the STM images [9.169]. Each molecule was adsorbed with the ring system parallel to a single-layer cluster of thirteen atoms. The energy of the system was minimized as a function of the adsorbate-metal separation for several molecular binding sites and orientations. At minimum energy separation, the electron density for occupied states (ρE) and the hole density for unoccupied states (ρH) were calculated at a given height above the plane of the molecule. All orbitals within either 1 eV above or below E_F were summed, and the images were convolved with a Gaussian of 1 Å width to simulate effects of a finite tunneling tip. Lowest Unoccupied Molecular Orbitals (LUMOs) and Highest Occupied Molecular Orbitals (HOMOs) for isolated molecules were also compared with the experimental images. The calculated results for three isomers of monomethylazulene on Pt(111) are depicted with the experimental data in Fig.9.9d,e. The LUMOs for the isolated molecules are quite similar for these three molecules to that of unsubstituted azulene. The ρH plots for the three molecules show notable density of states at the methyl substitution sites for all three molecules, and the calculated shapes of the molecules agree well with the experimental ones. Similarly good agreement between theory and experiment has been obtained for the other related molecules, with the exception of azulene, which probably rotates at room temperature so that the calculated asymmetry is not observed.

9.3.3 Benzene on Pt(111)

Isolated benzene molecules adsorbed on Pt(111) were observed by *Weiss* and *Eigler* [9.172] using an STM operating at 4 K. The benzene molecules appear as protrusions on the flat terraces of the metal substrate. The apparent shape of the molecules in STM images varies with the binding site on the surface. Three distinct types of images were observed for the isolated benzene molecules (left half of Fig.9.10): (1) a structure with three lobes similar to that previously observed for benzene on Rh(111) [9.191]; (2) a cylindrical volcano with a small depression in the center; and (3) a simple bump. By comparing these images with calculations by *Sautet* and *Bocquet* [9.175], to be described below, the different types of images were ascribed to different adsorption sites on the surface. The three-lobed structure was observed in two rotational orientations, 60° apart, and assigned to benzene in both hcp-type and fcc-type threefold hollow adsorption sites. The volcano structure, which was found only near other adsorbates or defects on the surface, was assigned to an on-top site, and the simple bump to a bridge site. The STM images provide useful information on the variation of the electronic structure with the chemical environment, while at the same time indicating caution is needed in identifying molecules on surfaces by STM.

Sautet and *Joachim* have used their Electron-Scattering Quantum Chemistry (ESQC) technique to calculate the tunneling current between the tip and substrate through a molecule from the generalized Landauer formula utilizing a scattering matrix which has been calculated exactly [9.173, 174]. Extended Hückel molecular-orbital theory was used to calculate the electron structure of the tip, adsorbate, and substrate. Excellent agreement was obtained between the theoretical calculations and previously published experiments for benzene on Rh(111) [9.173] and copper-phthalocyanine on Cu(100) [9.174]. The more recent study of STM images on Pt(111) yielded the different image shapes as a function of the binding site and type of adsorption (right half of Fig.9.10) [9.175]. *Ab initio* total-energy and electron calculations for benzene physisorbed on graphite and MoS_2 have also been reported recently showing the dependence of the calculated images on both tunneling voltage and adsorbate binding site [9.192].

9.3.4 Ethylene on Pt(111)

The adsorption of ethylene (C_2H_4) on Pt(111) as a function of temperature has been investigated with a variable temperature STM [9.176-178]. This system has widely been analysed in surface-science experiments by a wide variety of techniques [9.193, 195]. Chemical reactions in the system as a function of temperature have now been followed in STM images. Ethylene is known to adsorb with the C=C bond parallel to the Pt surface for temperatures up to 230 K [9.193, 194]. For 230K < T < 450K, ethylene

Benzene on Pt(111)

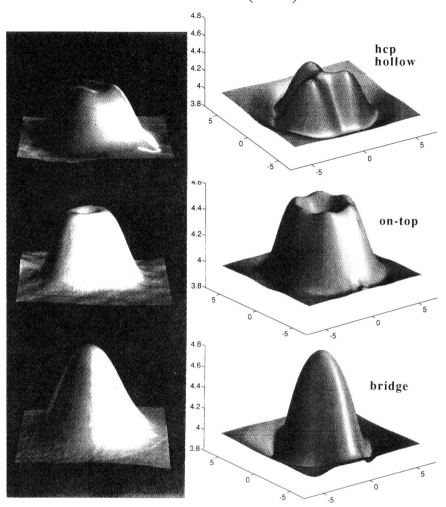

3 experimental STM images
P.S. Weiss, D.M. Eigler, IBM Almaden USA

Calculated STM images for 3 sites
M.-L. Bocquet, P. Sautet. CNRS Lyon France

Fig.9.10. Benzene on Pt(111), experiment (*left side*) and theory (*right side*). The experimental images correspond to three different 15×15 Å^2 regions of Pt(111), each showing a single adsorbed benzene molecule. The minimum to maximum height differences in the experimental images are (*top*) 0.58 Å, (*middle*) 0.72 Å, and (*bottom*) 0.91 Å, respectively. The calculated images are shown for the three labeled sites. (Experiment from [9.172], theory discussed in [9.175], figure courtesy of P. Sautet)

converts to ethylidyne (CCH_3) by loosing hydrogen, with ethylidyne bonded with the C-C bond perpendicular to the surface [9.199,33]. Further dehydrogenation of the ethylidyne leads to formation of "carbidic" carbon on the surface between 450 and 770 K [9.195]. Above 800 K, the carbon appears to convert to a graphitic structure. All steps in this reaction series have been observed in STM images.

For 160 K, individual structures of the size of ethylene molecules have been observed in an ordered pattern on the surface by STM [9.176]. Annealing the sample to 350 K leads to the formation of ethylidyne on the surface, but no structures are discernible by STM at this temperature. When the annealed surface is cooled to 180 K, however, STM images of the ethylidyne intermediate are obtained, showing both the ethylene, which has long-range order and a sharper pattern, and the ethylidyne, which appears rather disorderd with a fuzzier pattern [9.177]. Presumably the lower temperature allows the reduction of molecular motion, probably relatively low frequency vibrations, allowing the imaging of individual molecules by STM.

Conversion of ethylene to ethylidyne begins at 230 K, is nearly complete after $\approx$ 10 min at this temperature, and follows a first-order kinetics rate law in ethylene coverage [9.177]. *In situ* STM imaging of the conversion of ethylene to ethylidyne was performed. Figure 9.11 displays two images which were obtained during the reaction process [9.183]. Figure 9.11a corresponds to a partially-reacted surface, after the sample had been held for several minutes at 230 K, while Fig.9.11b was recorded several

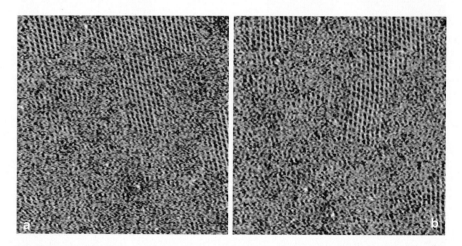

Fig.9.11a,b. Conversion of ethylene to ehtylidyne on Pt(111). These 400×400 Å^2 images were obtained as a function of time at 230 K. In these images, areas covered by ethylene molecules appear as the well-ordered pattern, and areas covered by ethylidyne molecules appear as a rather disordered pattern. (**a**) This image was obtained after annealing the sample for several minutes at 230 K. (**b**) This image was obtained on the same area of the surface kept for several additional minutes at the same temperature after the image shown in (**a**). From [9.177]

minutes later. Two small white protrusions in Fig.9.11a are shifted only slightly downward in Fig.9.11b and can be used as markers to examine changes in the images. The well-ordered sharper structures in the top center, upper-left, and middle-right edge of Fig.9.11a correspond to ethylene, which is parallel to the surface. The somewhat fuzzy, less ordered structures in the rest of the images correspond to ethylidyne, which is perpendicular to the surface. Comparison of the two images indicates that the reaction occurs at the edges of the ethylene islands, rather than randomly over the surface, as might have been expected from a simple model of first-order kinetics.

Further heating of the ethylidyne covered Pt(111) surface above 430 K results in further dehydrogenation and the observation of small carbonaceous particles on the surface in STM images [9.177]. The density of such particles on the surfaces increases with temperature from 430 to 500 K. Annealing the sample to 700 K and subsequent imaging at room temperature shows carbonaceous particles $10 \div 15$ Å in diameter uniformly distributed across the surface, with no tendency to cluster at step edges. Annealing to 800 K, room temperature imaging reveals graphite islands, $20 \div 30$ Å in diameter, uniformly distributed over the surface. Annealing to 1070 K shows graphite islands accumulating at lower step edges, along with many small islands still on the terraces [9.178]. Further annealing to 1230 K yields larger graphite islands at step edges and fewer large, regularly shaped islands on the terraces. Step edges decorated by graphite islands are much rougher than clean Pt step edges. Various superstructures with periods of 5 Å and $20 \div 22$ Å are observed and attributed to moiré-type patterns determined by the higher-order commensurability of the graphite and Pt lattices at different relative rotations. The graphite appears to be a single layer, and atomically resolved images show only three of the six carbon atoms in the graphite honeycomb structure. The observed structures are similar to those observed for bulk graphite, which had been attributed to the inequivalence of the A and B carbon atoms in the graphite lattice [9.196], resulting in the preferential imaging of B atoms in STM images due to electronic effects. Presumably the explanation is more complicated for the system of a single graphite layer on Pt(111), which may indicate that other factors play a role in the bulk graphite case as well.

9.3.5 C_{60} on Au(111) and Cu(111)

The buckminster-fullerene molecules, particularly C_{60}, have recently been the subject of many STM studies. The molecular arrangements and electronic structure of the overlayers on various substrates, such as Au(111) [9.179-182], Au(100) [9.183], Au(110) [9.184], Ag(111) [9.180,181], Cu(111) [9.185], GaAs(110) [9.186], Si(111) [9.187,188], and Si(100) [9.189], have been elucidated. Here we discuss in detail only two specific studies which observed the internal structure of the C_{60} molecule on Au(111) [9.181] and Cu(111) [9.185].

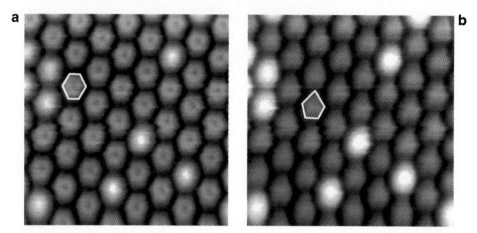

Fig.9.12. Simultaneously acquired (**a**) occupied ($-2\,$V) and (**b**) unoccupied ($1.5\,$V) state images of $(2\sqrt{3}\times2\sqrt{3})\text{R}30°$ C_{60} domains on Au(111) $(73\times73\,\text{Å}^2)$. From [9.181]

The interaction of C_{60} with Au(111) and Ag(111) has been investigated by STM in great detail by *Altman* and *Colton* [9.180,181]. Two close-packed adsorbate structures are observed for C_{60} on Au(111): (1) a layer with roughly 38×38 periodicity and (2) a layer with $(2\sqrt{3}\times2\sqrt{3})\text{R}30°$ unit cell in which all molecules are in equivalent surface sites [9.186]. Some molecules appear bright and some appear dim, and molecules change their contrast with time over a period of minutes [9.180]. The occurrence of such contrast was attributed to variations in the electronic structure as a result of different bonding to the substrate for molecules bound in different rotational orientations on the surface and to molecules rotating on the surface. Figure 9.12 shows that the observed internal structure of the molecules differs with tunneling voltage for simultaneously measured occupied and unoccupied state images of $(2\sqrt{3}\times2\sqrt{3})\text{R}30°$ C_{60} domains on Au (111) [9.181]. The occupied state image (Fig.9.12a) reveals sixfold molecular structures, each with a hole, while the unoccupied state image (Fig. 9.12b) appears to have pentagonal molecular shapes. Tunneling spectra indicate that there is a charge transfer from the metal to the molecule. The observed internal structure of C_{60} was attributed to the molecule being bonded in an on-top site with the five-membered ring on the Au(111) surface. Occupied state images may reflect the symmetry of the adsorption site, while unoccupied state images may reflect the symmetry of the molecular states.

C_{60} on the Cu(111)-(1×1) surface was observed using the Field-Ion Scanning Tunneling Microscope (FI-STM), and the experimental images compared well with the calculated charge density around the C_{60} molecules on a metal surface [9.185]. C_{60} forms a commensurate (4×4) monolayer when annealed to temperatures between 250 and 380° C. STM images of

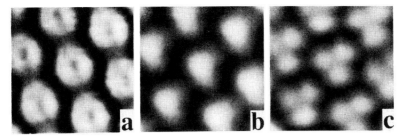

Fig.9.13a-c. 22×22 Å^2 images of C$_{60}$ molecules uniformly adsorbed on the Cu(111) surface at bias voltages of (**a**) -2.0 V, (**b**) -0.10 V, and (**c**) 2.0 V. STM images reveal a strong dependence on the bias voltage, but all exhibit the threefold symmetry. From [9.185]

C$_{60}$ on Cu(111) have a strong dependence on the bias voltage, but all indicate threefold molecular symmetry (Fig.9.13), suggesting that one of the hexagonal rings of the molecule faces down on the Cu(111) surface. The observation of four domains in the images and the alignment of one lobe of the molecule either along or 180° from the $<01\bar{1}>$ direction indicate a threefold hollow site for the molecule. Calculations of the voltage-dependent local charge density around the molecule agree well with the experimental data. This approach involves the calculation of a two-dimensional band structure for a fcc (111) layer of C$_{60}$ molecules using the local density approximation. The calculations indicate charge transfer from the Cu substrate to the C$_{60}$ monolayer.

9.3.6 Conclusions

It is clear that much progress has been made in the STM imaging of molecules on surfaces in a very short time. The new synergy of theory and experiment on these systems is particularly notable in elucidating the capabilities of the instrument for molecular identification. Molecular binding sites, orientations and diffusion rates are among the parameters which can be inferred from STM images. Real-space imaging has demonstrated the progress of chemical reactions and reactivity at steps and defects. Further progress is expected in understanding adsorbed molecules on surfaces and observing more complicated chemical reactions on an atomic scale.

9.4 STM on Superconductors[5]

9.4.1 STM on Conventional Superconductors

Since the impressive demonstration of the density of states near and in the vortex core by *Hess* et al. [9.197], many additional experiments have been performed:

Hess et al. [9.198] studied the vortex lattice as a function of the angle between the applied magnetic field and the basal plane. The observed lattice distortions at low angles, and the instabilities into ordered and buckled rows of vortices at angles exceeding 80° followed by a disordering transition at 85°. These transitions were explained by the anisotropy of the order parameter.

Renner et al. [9.199] investigated the $Nb_{1-x}Ta_xSe_2$ system. It was shown that the zero-bias anomaly observed at the vortex-core center in the pure $NbSe_2$ material disappears when Nb was replaced by Ta at about 20 wt.%. The data was explained by a transition from the clean to the dirty limit.

The choice of a suitable superconducting material for low-temperature scanning tunneling microscopy/spectroscopy is still rather restricted. The ideal nature of the $NbSe_2$ plays an important role in the success of the vacuum tunneling investigations for which it has been used. The van der Waals layers exposed upon cleaving are inert enough to allow good tunneling even after exposure to air. Additional favorable parameters are the reasonably high superconducting transition temperature and the possibility to grow large single crystals where atomically flat plateaux are obtained, simply by cleaving with an adhesive tape. However, *Kashiwaya* et al. [9.200] imaged the vortex lattice in NbC_xN_{1-x}. They claimed to have observed a particular vortex core structure, different from the normal density of states or the zero bias conductance peak.

Berthe et al. [9.201] investigated the influence of a transport current on the vortex lattice, although the detection of moving vortices was not possible. However, the vortex lattice disappeared when a critical transport current was reached.

Berthe et al. [9.202] and *Behler* et al. [9.203] studied the influence of ferromagnetic STM tip on the vortex lattice of $NbSe_2$. *Behler* et al. have shown that the inter-vortex distance is influenced by the magnetic stray field of the ferromagnetic probe. In zero external field a hexagonal vortex lattice with an inter-vortex distance corresponding to an external field of 15 mT was observed. This field is much smaller than the expected stray field of a ferromagnetic tip with a single magnetic domain at its end. The interpretation is still controversial.

The pinning of vortices in a $NbSe_2$ single crystal with correlated defects, produced by heavy ion bombardment was studied by *Behler* et al.

---

[5] This section is based on notes provided by H.J. Güntherodt and H.J. Hug.

[9.204]. Similar to the macroscopic magnetization measurements [9.205], *Behler* et al. observed a strong hysteresis of the inter-vortex distance upon field variations. Further, the hexagonal vortex lattice was severely distorted by the irradiation-generated defects.

9.4.2 STM on High-T_c Superconductors

Several groups have tried to spatially resolve the superconducting state of high transition-temperature (high-T_c) superconductors. Although good and spatially resolved tunneling spectra have been obtained [9.206] the detection of single vortices has not been demonstrated yet and the nature of the superconducting gap is still controversial. The complex structure of HTSC materials, degraded surfaces, Coulomb blockade, Schottky barriers and a possible drop of the order parameter towards the surface, all detract from easily gained access to the parameters of superconductivity with the STM's advantage of high spatial resolution. Point-contact spectroscopy overcomes the above problems, but the interpretation of the results remains controversial. Furthermore, the high mobility of the vortices even at low temperatures and the small coherence length of the order of some Ångstroms combine to make a detection of single vortices by STM rather difficult.

9.4.3 Scanning Probe Microscopy on High-T_c Superconductors

In contrast, the detection of vortices by measuring their magnetic stray field seems feasible, since the penetration depth and thus the magnetic diameter of the vortices, is large (some hundreds of nanometers). Indeed, not only the classical bitter technique and Lorentz microscopy [9.207], but also Scanning Hall Probe Microscopy (SHPM) [9.208], scanning SQUID microscopy [9.209] and, recently, magnetic force microscopy [9.210] have successfully imaged single vortices.

References

9.1   R. Berndt, J.K. Gimzewski, R.R. Schlittler: Ultramicroscopy **42-44**, 528 (1992)
9.2   O. Züger, U. Dürig: Ultramicroscopy **42-44**, 520 (1992)
9.3   S. Rousset, S. Gauthier, O. Siboulet, J.C. Girard, S. de Cheveigne, M. Huerta-Garnica, W. Sacks, M. Belin, J. Klein: Ultramicroscopy **42-44**, 515 (1992)
9.4   Y. Hasegawa, P. Avouris: Phys. Rev. Lett. **71**, 1071 (1993)
9.5   M.F. Crommie, C.P. Lutz, D.M. Eigler: Science **262**, 218 (1993)
9.6   M. Schmid, H. Stadler, P. Varga: Phys. Rev. Lett. **70**, 1441 (1993)
9.7   C. Dunphy, P. Sautet, D.F. Ogletree, M.B. Salmeron: J. Vac. Sci. Technol. A **11**, 2145 (1993)
9.8   C. Dunphy, P. Sautet, D.F. Ogletree, O. Dabbousi, M.B. Salmeron: Phys. Rev. B **47**, 2320 (1993)

9.9   H. Brune, J. Wintterlin, R.J. Behm, G. Ertl: Phys. Rev. Lett. **68**, 624 (1992)
9.10   H. Brune, J. Wintterlin, J. Trost, G. Ertl, J. Wiechers, R.J. Behm: J. Chem. Phys. **99**, 2128 (1993)
9.11   P.S. Weiss, D.M. Eigler: Phys. Rev. Lett. **69**, 2240 (1992)
9.12   P. Sautet, J. Dunphy, D.F. Ogletree, M. Salmeron: Surf. Sci. **295**, 347 (1993)
9.13   P. Sautet, J.C. Dunphy, D.F. Ogletree, C. Joachim, M. Salmeron: Surf. Sci. **315**, 127 (1994)
9.14   J.C. Dunphy, D.F. Ogletree, M.B. Salmeron, P. Sautet, M.-L. Bocquet, C. Joachim: Ultramicroscopy **42-44**, 490 (1991)
9.15   D.M. Eigler, P.S. Weiss, E.K. Schweizer, N.D. Lang: Phys. Rev. Lett. **66**, 1189 (1991)
9.16   T.A. Land, T. Michely, R.J. Behm, J.C. Hemminger, G. Comsa: Appl. Phys. A **53**, 1 (1991)
9.17   T.A. Land, T. Michely, R.J. Behm, J.C. Hemminger, G. Comsa: J. Chem. Phys. **97**, 6774 (1992)
9.18   T.A. Land, T. Michely, R.J. Behm, J.C. Hemminger, G. Comsa: Surf. Sci. **264**, 261 (1992)
9.19   S. Yamazaki, M. Tanaka, S. Tanaka, M. Fuginami, R. Uemori, D. Fujita, T. Homma, M. Ono: J. Vac. Sci. Technol. B **9**, 883 (1991)
9.20   P. Zeppenfeld, C.P. Lutz, D.M. Eigler: Ultramicroscopy **42-44**, 128 (1992)
9.21   R. Koch, E. Schwarz, K. Schmidt, B. Burg, K. Christmann, K.H. Rieder: Phys. Rev. Lett. **71**, 1047 (1993)
9.22   J.G. Forbes, A.J. Gellman, J.C. Dunphy, M. Salmeron: Surf. Sci. **279**, 68 (1992)
9.23   R.Q. Hwang, D.M. Zeglinski, A.L. Vazquez-de-Parga, D.F. Ogletree, G.A. Somorjai, M. Salmeron, D.R. Denley: Phys. Rev. B **44**, 1914 (1991)
9.24   D.F. Ogletree, R.Q. Hwang, D.M. Zeglinski, A.L. Vazquez-de-Parga, G.A. Somorjai, M. Salmeron: J. Vac. Sci. Technol. B **9**, 886 (1991)
9.25   D. Heuer, T. Müller, G. Pfnür, U. Köhler: Surf. Sci. Lett. **297**, L61 (1993)
9.26   A. Partridge, G.J. Tatlock, F.M. Leibsle, C.F. Flipse, G. Hörmandinger, J.B. Pendry: Phys. Rev. B **48**, 8267 (1993)
9.27   D. Bürgler, G. Tarrach, T. Schaub, R. Wiesendanger, H.J. Güntherodt: Phys. Rev. B **47**, 9963 (1993)
9.28   V. Maurice, P. Marcus: Surf. Sci. **262**, L59 (1992)
9.29   J.C. Dunphy, P. Sautet, D.F. Ogletree, M.B. Salmeron: J. Vac. Sci. Technol. A **11**, 1975 (1993)
9.30   K. Motai, T. Hashizume, H. Lu, D. Jeon, T. Sakurai, H.W. Pickering: Appl. Surf. Sci. **67**, 246 (1993)
9.31   R. Vogel, H. Baltruschat: Far. Discuss. **94**, 317 (1992)
9.32   W. Haiss, J.K. Sass, X. Gao, M.J. Weaver: Surf. Sci. **274**, L593 (1992)
9.33   R. Vogel, H. Baltruschat: Surf. Sci. **259**, L739 (1991)
9.34   R. Vogel, H. Baltruschat: Ultramicroscopy **42-44**, 562 (1992)
9.35   D.M. Eigler, C.P. Lutz, W.E. Rudfe: Nature **352**, 600 (1991)
9.36   H.C. Galloway, J.J. Benitez, M. Salmeron: Surf. Sci. **298**, 127 (1993)
9.37   J. Trost, J. Wintterlin, G. Ertl: Surf. Sci. (1994) submitted
9.38   L.P. Nielsen, F. Besenbacher, E. Laegsgaard, I. Steensgaard: Phys. Rev. B **44**, 13156 (1991)
9.39   H. Brune, J. Wintterlin, R.J. Behm, G. Ertl: Phys. Rev. B (1994) submitted
9.40   R. Schuster, J.V. Barth, G. Ertl, R.J. Behm: Phys. Rev. B **44**, 13689 (1991)
9.41   G. Doyen, D. Drakova, J.V. Barth, R. Schuster, T. Gritsch, R.J. Behm, G. Ertl: Phys. Rev. B **48**, 1738 (1993)
9.42   R. Schuster, J.V. Barth, G. Ertl, R.J. Behm: Phys. Rev. Lett. **69**, 2547 (1992)
9.43   J.V. Barth, R.J. Behm, G. Ertl: Surf. Sci. **302**, L319 (1994)

9.44 J.V. Barth, H. Brune, R. Schuster, G. Ertl, R.J. Behm: Surf. Sci. **292**, L769 (1993)

9.45 J.V. Barth, R. Schuster, G. Ertl, R.J. Behm: Surf. Sci. **302**, 158 (1994)

9.46 J.V. Barth, R. Schuster, J. Wintterlin, G. Ertl, R.J. Behm: Phys. Rev. B (1994) submitted

9.47 C. Klink, L. Olesen, F. Besenbacher, I. Steensgaard, E. Laegsgaard, N.D. Lang: Phys. Rev. Lett. **71**, 4350 (1993)

9.48 G. Hörmandinger, J.B. Pendry, F.M. Leibsle, P.W. Murray, R.W. Joyner, G. Thorton: Phys. Rev. B **48**, 8356 (1993)

9.49 O. Haase, R. Koch, M. Borbonus, K.H. Rieder: Ultramicroscopy **42-44**, 541 (1992)

9.50 F. Leibsle: Surf. Sci. **297**, 98 (1993)

9.51 F.M. Leibsle, C.F.J. Flipse, A.W. Robinson: Phys. Rev. B **47**, 15865 (1993)

9.52 M. Voetz, H. Niehus, J. O'Connor, G. Comsa: Surf. Sci. **292**, 211 (1993)

9.53 H. Niehus, R. Spitzl, K. Besocke, G. Comsa: Phys. Rev. B **43**, 12619 (1991)

9.54 F.M. Leibsle, R. Davis, A.W. Robinson: Phys. Rev. B **49**, 8290 (1994)

9.55 F.M. Leibsle: Surf. Sci. **311**, 45 (1994)

9.56 F. Jensen, F. Besenbacher, E. Laegsgaard, I. Stensgaard: Surf. Sci. **259**, L774 (1991)

9.57 F. Jensen, F. Besenbacher, I. Stensgaard: Surf. Sci. **269/270**, 400 (1992)

9.58 F. Jensen, F. Besenbacher, E. Laegsgaard, I. Stensgaard: Phys. Rev. B **42**, 9206 (1990)

9.59 C. Wöll, R.J. Wilson, S. Chiang, H.C. Zeng, K.A.R. Mitchell: Phys. Rev. B **42**, 11926 (1990)

9.60 L. Eierdal, F. Besenbacher, E. Laegsgaard, I. Stensgaard: Surf. Sci. **312**, 31 (1994)

9.61 O. Haase, R. Koch, M. Borbonus, K.H. Rieder: Phys. Rev. Lett. **66**, 1725 (1991)

9.62 O. Haase, R. Koch, M. Borbonus, K.H. Rieder: Chem. Phys. Lett. **190**, 621 (1992)

9.63 R. Koch, O. Haase, M. Borbonus, K.H. Rieder: Surf. Sci. **272**, 17 (1992)

9.64 L. Ruan, F. Besenbacher, I. Stensgaard, E. Laegsgaard: Phys. Rev. Lett. **70**, 4079 (1993)

9.65 K. Kern, H. Niehus, A. Schatz, P. Zeppenfeld, J. Goerge, G. Comsa: Phys. Rev. Lett. **67**, 855 (1991)

9.66 T. Hashizume, M. Taniguchi, K. Motai, H. Lu, K. Tanaka, T. Sakurai: Surf. Sci. **266**, 282 (1992)

9.67 T. Hashizume, M. Taniguchi, K. Motai, H. Lu, K. Tanaka, T. Sakurai: Ultramicroscopy **42-44**, 553 (1992)

9.68 M. Taniguchi, K. Tanaka, T. Hashizume, T. Sakurai: Surf. Sci. **262**, L123 (1992)

9.69 T. Schimizu, M. Tsukada: Solid State Commun. **87**, 193 (1993)

9.70 V.R. Dhanak, K.C. Prince, R. Rosei, P.W. Murray, F.M. Leibsle, M. Bowker, G. Thornton: Phys. Rev. B **49**, 5585 (1994)

9.71 P.W. Murray, F.M. Leibsle, Y. Li, Q. Guo, M. Bowker, G. Thornton, V.R. Dhanak, K.C. Prince, R. Rosei: Phys. Rev. B **47**, 12976 (1993)

9.72 A. Meyer, Y. Kuk, P.J. Estrup, P.J. Silverman: Phys. Rev. B **44**, 9104 (1991)

9.73 T. Wiederholt, H. Brune, J. Wintterlin, R.J. Behm, G. Ertl: Surf. Sci. (1994) submitted

9.74 L. Ruan, I. Stensgaard, F. Besenbacher, E. Laegsgaard: Phys. Rev. Lett. **71**, 2963 (1993)

9.75 L. Ruan, I. Stensgaard, F. Besenbacher, E. Laegsgaard: Ultramicroscopy **42-44**, 498 (1992)

9.76 L. Ruan, I. Stensgaard, E. Laegsgaard, F. Besenbacher: Surf. Sci. **296**, 275 (1993)

9.77 J.S. Ozcomert, W.W. Pai, N.C. Bartelt, J.E. Reutt-Robey: Phys. Rev. Lett. **72**, 258 (1994)

9.78 E. Kopatski, R.J. Behm: Phys. Rev. Lett. (1994) submitted

9.79 L. Ruan, F. Besenbacher, I. Stensgaard, E. Laegsgaard: Phys. Rev. Lett. **69**, 3523 (1992)

9.80 P. Murray, G. Thornton, M. Bowker, V.R. Dhanak, A. Baraldi, R. Rosei, M. Kiskinova: Phys. Rev. Lett. **71**, 4369 (1993)

9.81 F.M. Leibsle, S.M. Francis, R. Davis, N. Xiang, S. Haq, M. Bowker: Phys. Rev. Lett. **72**, 2569 (1994)

9.82 S.M. Francis, F.M. Leibsle, S. Haq, N. Xiang, M. Bowker: Surf. Sci. **315**, 284 (1994)

9.83 F.M. Leibsle, P.W. Murray, S.M. Francis, G. Thornton, M. Bowker: Nature **363**, 706 (1993)

9.84 R. Schuster, J.V. Barth, J. Wintterlin, R.J. Behm, G. Ertl: Phys. Rev. B (1994) submitted

9.85 L. Ruan, I. Stensgaard, E. Laegsgaard, F. Besenbacher: Surf. Sci. **314**, L873 (1994)

9.86 S.-L. Chang, P.A. Thiel: Crit. Rev. **3**, 239 (1994)

9.87 M. Copel, M.C. Reuter, E. Kaxiras, R.M. Tromp: Phys. Rev. Lett. **63**, 632 (1989)

9.88 E. Bauer: Z. Krist. **110**, 372 (1958)

9.89 J. Vrijmoeth, H.A. van der Vegt, J.A. Meyer, E. Vlieg, R.J. Behm: Phys. Rev. Lett. **72**, 3843 (1994)

9.90 J.A. Meyer, R.J. Behm: Phys. Rev. Lett. **73**, 364 (1994)

9.91 J.A. Meyer, J. Vrijmoeth, H.A. van der Vegt, E. Vlieg, R.J. Behm: Submitted

9.92 C. Günther, S. Günther, E. Kopatzki, R.Q. Hwang, J. Schröder, J. Vrijmoeth, R.J. Behm: Ber. Bunsenges. Phys. Chem. **97**, 522 (1993)

9.93 S. Günther, E. Kopatzki, M.C. Bartelt, J.W. Evans, R.J. Behm: Phys. Rev. Lett. **73**, 553 (1994)

9.94 R.Q. Hwang, C. Günther, J. Schröder, S. Günther, E. Kopatzki, R.J. Behm: J. Vac. Sci. Technol. A **10**, 1970 (1992)

9.95 S. Günther, A. Hitzke, R.J. Behm: To be published

9.96 C.A. Lang, M.M. Dovek, J. Nogami, C.F. Quate: Surf. Sci. **224**, L947 (1989)

9.97 D.D. Chambliss, R.J. Wilson, S. Chiang: J. Vac. Sci. Technol. **9**, 933 (1991)

9.98 J.A. Stroscio, D.T. Pierce, R.A. Dragoset: Phys. Rev. Lett. **70**, 3615 (1993)

9.99 J.A. Stroscio, D.T. Pierce: J. Vac. Sci. Technol. B **12**, 1783 (1994)

9.100 J.A. Stroscio, D.T. Pierce: Phys. Rev. B **49**, 8522 (1994)

9.101 E. Kopatzki, S. Günther, W. Nichtl-Pecher, R.J. Behm: Surf. Sci. **284**, 154 (1993)

9.102 M. Bott, T. Michely, G. Comsa: Surf. Sci. **272** 161 (1992)

9.103 M. Bott, M. Hohage, T. Michely, G. Comsa: Phys. Rev. Let. **70**, 1489 (1993)

9.104 T. Michely, M. Hohage, M. Bott, G. Comsa: Phys. Rev. Lett. **70**, 3943 (1993)

9.105 S.L. Chang, S. Günther, P.A. Thiel, R.J. Behm: to be published

9.106 S. Rousset, S. Chiang, D.E. Fowler, D.D. Chambliss: Phys. Rev. Lett. **69**, 3200 (1992)

9.107 S. Rousset, S. Chiang, D.E. Fowler, D.D. Chambliss: Surf. Sci. **287/288**, 941 (1993)

9.108 J.A. Meyer, R.J. Behm: To be published

9.109 S. Günther, R.J. Behm: To be published

9.110 A. Hitzke, S. Günther, R.J. Behm: To be published

9.111 M.M. Dovek, C.A. Lang, J. Nogami, C.F. Quate: Phys. Rev. B **40**, 11973 (1989)

9.112 D.D. Chambliss, R.J. Wilson: J. Vac. Sci. Technol. **9**, 928 (1991)

9.113 J. Wollschläger, N.M. Amer: Surf. Sci. **277**, 1 (1992)

9.114 B. Voigtländer, G. Meyer, N.M. Amer: Phys. Rev. B **44**, 10354 (1991)

9.115 B. Voigtländer, G. Meyer, N.M. Amer: Surf. Sci. **255**, L529 (1991)

9.116 J.A. Stroscio, D.T. Pierce, R.A. Dragoset, P.N. First: J. Vac. Sci. Technol. A **10**, 1981 (1992)

9.117 S. Chiang, D.D. Chambliss, V.M. Hallmark, R.J. Wilson, C. Wöll: In *The Structure of Surfaces III*, ed. by S.Y. Tong, M.A. Van Hove, K. Takayanagi, X.D. Xie, Springer Ser. Surf. Sci., Vol.24 (Springer, Berlin, Heidelberg 1991) p.204

9.118 D.D. Chambliss, R.J. Wilson, S. Chiang: Phys. Rev. Lett. **66**, 1721 (1991)

9.119 E.I. Altman, R.J. Colton: Surf. Sci. **304**, L400 (1994)

9.120 D.D. Chambliss, S. Chiang: Surf. Sci. **264**, L187 (1992)

9.121 D.D. Chambliss, R.J. Wilson, S. Chiang: J. Vac. Sci. Technol. A **10**, 1993 (1992)

9.122 A.K. Schmid, J. Kirschner: Ultramicroscopy **42-44**, 483 (1992)

9.123 A.K. Schmid, D. Atlan, H. Itoh, B. Heinrich, T. Ichinokawa, J. Kirschner: Phys. Rev. B **48**, 2855 (1993)

9.124 D.D. Chambliss, K.E. Johnson: Surf. Sci. **313**, 215 (1994)

9.125 K.E. Johnson, D.D. Chambliss, R.J. Wilson, S. Chiang: J. Vac. Sci. Technol. A **11**, 1654 (1993)

9.126 K. Kalki, D.D. Chambliss, K.E. Johnson, R.J. Wilson, S. Chiang: Phys. Rev. B **48**, 18344 (1993)

9.127 D.D. Chambliss, K.E. Johnson, K. Kalki, S. Chiang, R.J. Wilson: MRS Proc. **313**, 713 (1993)

9.128 M. Wuttig, B. Feldmann, J. Thomassen, F. May, H. Zillgen, A. Brodde, H. Hannemann, H. Neddermeyer: Surf. Sci. **291** 14 (1993)

9.129 A. Brodde, H. Neddermeyer: Surf. Sci. **287/288**, 988 (1993)

9.130 J. Giergiel, J. Kirschner, J. Landgrad, J. Shen, J. Woltersdorf: Surf. Sci. **310**, 1 (1994)

9.131 K.E. Johnson, D.D. Chamnbliss, R.J. Wilson, S. Chiang: Surf. Sci. **313**, L811 (1994)

9.132 Y. Gauthier, M. Poensgen, M. Wuttig: Surf. Sci. **303**, 36 (1994)

9.133 W.E. McMahon, E.S. Hirschorn, T.-C. Chiang: Surf. Sci. **279**, L231 (1992)

9.134 J. de la Figuera, J.E. Prieto, C. Ocal, R. Miranda: Phys. Rev. B **47**, 13043 (1993)

9.135 J. de la Figuera, J.E. Prieto, C. Ocal, R. Miranda: Surf. Sci. **307-309**, 538 (1994)

9.136 A. Brodde, H. Neddermeyer: Ultramicroscopy **42-44**, 556 (1992)

9.137 J.A. Stroscio, D.T. Pierce, J. Unguris, R.J. Celotta: J. Vac. Sci. Technol. B **12**, 1789 (1994)

9.138 D.T. Pierce, J.A. Stroscio, J. Unguris, R.J. Celotta: Phys. Rev. B **49**, 14564 (1994)

9.139 A. Brodde, G. Wilhelmi, D. Badt, H. Wengelnik, H. Neddermeyer: J. Vac. Sci. Techol. **9**, 920 (1991)

9.140 L. Pleth Nielsen, F. Besenbacher, I. Stensgaard, E. Laegsgaard: Phys. Rev. Lett. **71**, 754 (1993)

9.141 L. Pleth Nielsen, F. Besenbacher, E. Laegsgaard, I. Stensgaard: To be published

9.142 H. Röder, E. Hahn, H. Brune, J.-P. Bucher, K. Kern: Nature **366**, 141 (1993)

9.143 E. Hahn, E. Kampshoff, K. Kern: Chem. Phys. Lett. **223**, 347 (1994)

9.144 H. Röder, H. Brune, J.-P. Bucher, K. Kern: Surf. Sci. **298**, 121 (1993)

9.145 H. Röder, R. Schuster, H. Brune, K. Kern: Phys. Rev. Lett. **71**, 2086 (1993)

9.146 H. Brune, H. Röder, C. Boragno, K. Kern: Phys. Rev. B **49**, 2997 (1994)

9.147 P. Grütter, U.T. Dürig: J. Vac. Sci. Technol. B **12**, 1768 (1994)

9.148 P. Grütter, U.T. Dürig: Phys. Rev. B **49**, 2021 (1994)

9.149 R.Q. Hwang: to be published

9.150 R.Q. Hwang, J. Schröder, C. Günther, R.J. Behm: Phys. Rev. Lett. **67**, 3279 (1991)

9.151 R.Q. Hwang, R.J. Behm: J. Vac. Sci. Technol. B **10**, 256 (1992)

9.152 G.O. Pötschke, J. Schröder, C. Günther, R.Q. Hwang, R.J. Behm: Surf. Sci. **251/252**, 592 (1991)

9.153 J. Schröder, C. Günther, R.Q. Hwang, R.J. Behm: Ultramicroscopy **42-44**, 475 (1992)

9.154 J. Vrijmoeth, C. Günther, J. Schröder, R.Q. Hwang, R.J. Behm: In *Magnetism and Structure in Systems of Reduced Dimension*, ed. by R.F.C. Farrow et al. (Plenum, New York 1993) p.55

9.155 G.O. Pötschke, R.J. Behm: Phys. Rev. B **44**, 1442 (1991)

9.156 C. Günther, J. Vrijmoeth, R.Q. Hwang, R.J. Behm: Submitted

9.157 T.E. Madey, K.-J. Song, C.-Z. Dong, R.A. Denimm: Surf. Sci. **247**, 175-187 (1991)

9.158 K.-J. Song, R.A. Demmin, C. Dong, E. Garfunkel, T.E. Madey: Surf. Sci. **227**, L79 (1990)

9.159 S. Esch, M. Hohage, T. Michely, G. Comsa: Phys. Rev. Lett. **72**, 518 (1994)

9.160 V. Renz, C. Günther, J.A. Meyer, R.J. Behm: To be published

9.161 J.A. Venables, G.D. Spiller, M. Hanbücken: Rep. Prog. Phys. **47**, 399 (1984)

9.162 Further references can be found in the respective experimental papers

9.163 M. Bäumer, D. Cappus, G. Illing, H. Kuhlenbeck, H.-J. Freund: J. Vac. Sci. Technol. A **10**, 2407 (1992)

9.164 C.A. Ventrice, Jr., T. Bertrams, H. Hannemann, A. Brodde, H. Neddermeyer: Phys. Rev. B **49**, 5773 (1994)

9.165 P. Schmid, J.A. Meyer, R.J. Behm: To be published

9.166 H.C. Galloway, J.J. Benitez, M. Salmeron: Surf. Sci. **298**, 127 (1993)

9.167 H.-C. Wang, D.F. Ogletree, M. Salmeron: J. Vac. Sci. Technol. B **9**, 853 (1991)

9.168 V.M. Hallmark, S. Chiang: Surf. Sci. **286**, 190 (1993)

9.169 V.M. Hallmark, S. Chiang, K.-P. Meinhardt, K. Hafner: Phys. Rev. Lett. **70**, 3740 (1993)

9.170 V.M. Hallmark, S. Chiang, J.K. Brown, C. Woell: In *Synthetic Microstructures in Biological Research*, ed. by J.M. Schnur, M. Peckerar (Plenum, New York 1992)

9.171 S. Chiang, V.M. Hallmark, K.-P. Meinhardt, K. Hafner: J. Vac. Sci. Technol. B **12**, 1957 (1994)

9.172 P.S. Weiss, D.M. Eigler: Phys. Rev. Lett. **71**, 3139 (1993)

9.173 P. Sautet, C. Joachim: Chem. Phys. Lett. **185**, 23 (1991)

9.174 P. Sautet, C. Joachim: Surf. Sci. **271**, 387 (1992)

9.175 P. Sautet, M.-L. Bocquet: Surf. Sci. Lett. **304**, L445 (1994)

9.176 T.A. Land, T. Michely, R.J. Behm, J.C. Hemminger, G. Comsa: Appl. Phys. A **53**, 414 (1991)

9.177 T.A. Land, T. Michely, R.J. Behm, J.C. Hemminger, G. Comsa: J. Chem. Phys. **97**, 6774 (1992)

9.178 T.A. Land, T. Michely, R.J. Behm, J.C. Hemminger, G. Comsa: Surf. Sci. **264**, 261 (1992)

9.179 R.J. Wilson, G. Meijer, D.S. Bethune, R.D. Johnson, D.D. Chambliss, M.S. deVries, H.E. Hunziker, H.R. Wendt: Nature **348**, 621 (1990)

9.180 E.I. Altman, R.J. Colton: Surf. Sci. **279**, 49 (1992); ibid. **295**, 13 (1993); and in *Atomic and Nanoscale Modification of Materials: Fundamentals and Applica-*

tions, Vol.E239 of NATO Advanced Study Institute, Series E: Applied Science, ed. by P. Avouris (Plenum, New York 1993) p.303

9.181 E.I. Altman, R.J. Colton: Phys. Rev. B **48**, 18244 (1993)

9.182 H.P. Lang, V. Thommen-Geiser, C. Bolm, M. Felder, J. Frommer, R. Wiesendanger, H. Werner, R. Schlögl, A. Zahab, P. Bernier, G. Gerth, D. Anselmetti, H.-J. Guntherodt: Appl. Phys. A **56**, 197 (1993)

9.183 Y. Kuk, D.K. Kim, Y.D. Suh, K.H. Park, H.P. Noh, S.J. Ohn, S.K. Kim: Phys. Rev. Lett. **70**, 1948 (1993)

9.184 R. Gaisch, R. Berndt, J.K. Gimzewski, B. Reihl, R.R. Schlittler, W.D. Schneider, M. Tschudy: Appl. Phys. A **57**, 207 (1993)

9.185 T. Hashizume, K. Motai, X.D. Wang, H. Shinohara, Y. Saito, Y. Maruyama, K. Ohno, Y. Kawazoe, Y. Nishina, H.W. Pickering, Y. Kuk, T. Sakurai: Phys. Rev. Lett. **71**, 2959 (1993)

9.186 Y.Z. Li, J.C. Patrin, M. Chander, J.H. Weaver, L.P.F. Chibante, R.E. Smalley: Science **252**, 547 (1991)

9.187 X.-D. Wang, T. Hashizume, H. Shinohara, Y. Saito, Y. Nishina, T. Sakurai: Jpn. J. Appl. Phys. **31**, L983 (1992)

9.188 H. Xu, D.M. Chen, W.N. Creager: Phys. Rev. Lett. **70**, 1850 (1993)

9.189 T. Hashizume, S.-D. Wang, Y. Nishina, H. Shinohara, Y. Saito, Y. Kuk, T. Sakurai: Jpn. J. Appl. Phys. **31**, L880 (1992)

9.190 V.M. Hallmark, S. Chiang, J.K. Brown, C. Woell: Phys. Rev. Lett. **66**, 48 (1991)

9.191 H. Ohtani, R.J. Wilson, S. Chiang, C.M. Mate: Phys. Rev. Lett. **60** 2398 (1988)

9.192 A.J. Fisher, P.E. Blöchl: Phys. Rev. Lett. **70**, 3263 (1993)

9.193 R.J. Koestner, J. Stöhr, J.L. Gland, J.A. Hosley: Chem. Phys. Lett. **105**, 333 (1984)

9.194 H. Steininger, H. Ibach, S. Lehwald: Surf. Sci. **117**, 685 (1982)

9.195 C.L. Pettiette-Hall, D.P. Land, R.T. McIver, J.C. Hemminger: J. Phys. Chem. **94**, 1948 (1990)

9.196 I.P. Batra, N. Garcia, H. Rohrer, H. Salemink, E. Stoll, S. Ciraci: Surf. Sci. **181**, 126 (1987)

9.197 H.F. Hess, R.B. Robinson, R.C. Dynes, J.M. Valles, J.V. Wasczak: Phys. Rev. Lett. **62**, 214 (1989)

9.198 H.F. Hess, C.A. Murray, J.V. Wasczak: Phys. Rev. Lett. **64**, 2138 (1992)

9.199 C. Renner, A.D. Kent, P. Niedermann, O. Fischer, F. Levy: Phys. Rev. Lett. **67**, 1650 (1991)

9.200 S. Kashiwaya, M. Koyanagi, A. Shoji: Appl. Phys. Lett. **61**, 1847 (1992)

9.201 R. Berthe, U. Hartmann, C. Heiden: Ultramicroscopy **42-44**, 696 (1992)

9.202 R. Berthe, U. Hartmann, C. Heiden: Appl. Phys. Lett. **57**, 2351 (1990)

9.203 S. Behler, S.H. Pan, M. Bernasconi, P. Jess, H.J. Hug, O. Fritz, H.-J. Güntherodt: J. Vac. Sci. Techn. B **12**, 2209 (1994)

9.204 S. Behler, S.H. Pan, P. Jess, A. Baratoff, H.-J. Güntherodt, F. Levy, G. Wirth, J. Wiesener: Phys. Rev. Lett. **72**, 1750 (1994)

9.205 P. Bauer, C. Giethmann, M. Kraus, T. Marek, J. Burger, G. Kreiselmeyer, G. Saemann-Ischenko, M. Skibowski: Europhys. Lett. **53**, 585 (1993)

9.206 C. Renner: Low temperature scanning tunneling microscopy and spectroscopy of layered superconductors. Dissertation, University of Geneva (1993)

9.207 K. Haralda, T. Matsuda, H. Kasai, J.E. Bonevich, T. Yoshida, U. Kawabe, A. Tonomura: Phys. Rev. Lett. **71**, 3371 (1993)

9.208 A.M. Chang, H.D. Hallen, H.F. Hess, H.L. Kao, J. Kwo, A. Sudbo, T.Y. Chang: Europ. Phys. Lett. **20**, 645 (1992)
A.M. Chang, H.D. Hallen, L. Harriot, H.F. Hess, H.L. Kao, R.E. Miller, R. Wolfe, J. van der Ziel: APL **61**, 1974 (1992)

9.209 C.C. Tsuei, J.R. Kirtley, C.C. Chi, L.S. Yu-Jahnes, A. Gupta, T. Shaw, J.Z. Sun, M.B. Ketchen: Phys. Rev. Lett. (submitted)
9.210 A. Moser, H.J. Hug, I. Parashikov, B. Stiefel, O. Fritz, H. Thomas, H.-J. Güntherodt: Phys. Rev. Lett. (submitted)
H.J. Hug, A. Moser, I. Parashikov, B. Stiefel, O. Fritz, H. Thomas, H.-J. Güntherodt: Physica C, to be published
H.J. Hug, A. Moser, I. Parashikov, B. Stiefel, O. Fritz, H. Thomas, H.-J. Güntherodt: NATO ASI (1994), to be published

Subject Index